Field Guide to
Coastal Wetland Plants
of the Southeastern
United States

Coastal marshes in the Santee River Delta in South Carolina

Field Guide to Coastal Wetland Plants of the Southeastern United States

RALPH W. TINER

Drawings by Abigail Rorer

The University of Massachusetts Press
AMHERST

Copyright © 1993 by Ralph W. Tiner
All rights reserved
Printed in the United States of America
LC 92–36526
ISBN 0–87023–832-9 (cloth); 833–7 (pbk.)
Set in Sabon by Keystone Typesetting, Inc.
Printed and bound by Thomson-Shore, Inc.

Library of Congress Cataloging-in-Publication Data
Tiner, Ralph W.
    Field guide to coastal wetland plants of the southeastern United States /
Ralph W. Tiner ; drawings by Abigail Rorer.
        p.   cm.
    Includes bibliographical references (p.) and index.
    ISBN 0–87023–832–9 (cloth : alk. paper).
    ISBN 0–87023–833–7(pbk. : alk. paper)
    1. Wetland flora—Southern States—Identification.
    2. Coastal flora—Southern States—Identification.
    3. Wetland flora—Southern States—Pictorial works.
    4. Coastal flora—Southern States—Pictorial works. I. Title.
QK125.T55   1993
582. 0975—dc20   92–36526
                                                    CIP

British Library Cataloguing in Publication data are available.

To
John S. Rankin, Jr.,
in memoriam,
for introducing me to wetlands

Contents

Preface

It has been more than two decades since my first in-depth encounter with coastal wetlands. As a graduate student at the University of Connecticut, my first day on the job as a member of a tidal marsh survey field team was a day to remember. In fact, it was one of my most difficult days in the wetlands. It was a hot, humid summer day and the place was a tidal cattail marsh in Connecticut. With my first step into the marsh I sank nearly to my waist! My lighter-footed, more experienced colleagues enjoyed the show as the former football player pulled one leg out of the marsh ooze, only to drive the other leg deeper into the muck. A most embarrassing and unpleasant experience—yet, despite this rude introduction, I somehow persevered and now have spent my entire professional life in wetlands. I guess I really never got out of the quagmire.

After graduate school and mapping Connecticut's tidal wetlands, I accepted a position as a marine biologist with the South Carolina Wildlife and Marine Resources Department in Charleston. It was here that I first sampled some of the productivity of southeastern coastal marshes (raw oysters, boiled shrimp, and steamed crabs) and observed firsthand the richness of their flora. Although many salt marsh species were the same as those in Connecticut, many were quite different and unique to the Southeast.

In particular, the tidal fresh marshes were extremely diverse in plant composition. In fact, some of these marshes are among the most diverse plant communities in the country, outside the tropics, and are, in my opinion, vital areas for the preservation of biodiversity. While working in South Carolina and enjoying the rich flora of its coastal marshes, I often contemplated preparing a field guide of coastal marsh plants to aid nonbotanists in their identification. Before leaving South Carolina, I did manage to assemble a report on their coastal marshes which, among other things, included habitat descriptions and a list of common plants of coastal marshes and impoundments.

Although there are a few general booklets for identifying dominant coastal wetland plants in some states, there is no comprehensive, easy-to-use field guide for identifying the majority of plants encountered in southeastern coastal marshes and mangrove swamps. The recent success of my two wetland field guides for the Northeast encouraged me to prepare one for the Southeast. The list of plant species included in this book came from my own observations plus a review of significant ecological studies of southern coastal wetlands. I have attempted to make this book as complete as possible but have undoubtedly missed some species. I intentionally did not cover in

great detail plants of mangrove swamps because tropical flora are unique to South
Florida and worthy of a separate volume,
but the major species are represented. In
addition, for tidal fresh marshes and
swamps, I attempted to include the more
common species, but since limited information on their plant composition exists, my
listing is probably conservative. Still, this
book includes descriptions and illustrations
of more than 250 species and additional
references to more than 200 other species.
Plants not included in this guide can be
identified by using any of a number of field
guides to wildflowers or trees and shrubs or
by consulting more technical botanical
manuals.

This field guide is designed to help those
with little or no training in botany to identify coastal wetland plants in the Southeast.
It will be useful to general biologists and
ecologists, park naturalists, environmental
engineers, landscape architects, planners,
developers, teachers and students in environmental sciences, and others interested in
wetlands. The field guide, while emphasiz-
ing plant identification, also includes dis-
cussions of the ecology of southern coastal
wetlands and their distribution to give the
reader a better understanding of these valuable natural resources. I hope that an increased knowledge of these wetlands will
stimulate an interest in their conservation
to insure that their functions and values are
protected not only for ourselves but also
for future generations.

Acknowledgments

I wish to thank several individuals who made significant contributions to the completion of this book. Most importantly, Abbie Rorer brought this book to life with her fine plant illustrations, many of which have been used in my other books: *Field Guide to Coastal Wetland Plants of the Northeastern United States* and *Field Guide to Nontidal Wetland Identification*. Roberta Lombardi and Claire Johnson provided access to plant specimens from the University of Massachusetts Herbarium. John Hefner, Becky Stanley, and Don Woodard of the U.S. Fish and Wildlife Service were among numerous persons providing useful background material. Don Field of the National Oceanic and Atmospheric Administration contributed information on the acreage of coastal wetlands. Bill Niering of Connecticut College reviewed the final draft manuscript and provided helpful comments. Sue King and Cathy Zezima typed portions of the manuscript, which was coordinated by my wife, Barbara. I am grateful for Barbara's support throughout this project, especially for her indulgence of piles of my manuscript drafts and reference materials in our dining room. I also wish to acknowledge the efforts of the University of Massachusetts Press, especially Pam Wilkinson, Jack Harrison, and Barbara Palmer, in preparing this book for publication.

Besides the above persons, I would like to extend my gratitude to three individuals with whom I had the pleasure of working in South Carolina more than fifteen years ago: Mike MacKenzie for the opportunity to work in South Carolina's coastal marshes and Rob Dunlap and Curt Laffin for their assistance in the field and their efforts in helping protect these valuable wetlands.

Introduction

Tidal wetlands are a dominant landscape feature of the coastal zone in the southeastern United States. These wetlands are largely colonized by salt-tolerant plants forming salt and brackish marshes or mangrove swamps, or by other water-tolerant species in tidally influenced freshwater areas. Anyone driving to southern beaches is likely to have crossed coastal wetlands on the trip, since these wetlands often stretch for miles between barrier island beaches and the mainland. Oftentimes, people, both young and old, can be seen fishing or crabbing from bridges crossing tidal rivers or creeks. This provides direct evidence of the productivity and value of these natural habitats.

The most productive of our southern coastal marshes produce over 10 tons of organic matter per acre each year. This amount exceeds the yield of many American cornfields. Coastal wetlands can be viewed as aquatic farmlands. The organic matter produced supports a complex food chain of animals in estuarine and nearshore marine waters. The organic matter comes mainly from the leaves and stems of herbaceous marsh plants and also from mangrove leaves (in Florida). Each fall, when the herbs die back, their leaves and stems fall to the ground or in the water, where they are gradually broken down by various organisms into small fragments called *detritus*. This detritus provides food for numerous microorganisms (e.g., zooplankton), macroinvertebrates (e.g., marine worms, amphipods, shrimp, and crabs), and forage fishes (e.g., killifish, silversides, mullet, spot, and menhaden), which in turn are the food for larger fishes, such as flounder, bluefish, and weakfish (sea trout). These fishes as well as crabs and shrimp provide food for humans, which completes the food chain between coastal wetlands and mankind.

The vital link between coastal wetlands and adjacent waters makes these estuarine habitats critical nursery and spawning grounds for the majority of the nation's important recreational and commercial fish and shellfish species. Among these are menhaden, spot, mullet, flounder, croaker, redfish, weakfish, penaeid shrimp, blue crabs, and oysters. Over 90 percent of the Southeast's commercial fish landings come from estuarine-dependent fishes. The productivity of the Southeast's multimillion-dollar shrimp fishery is directly dependent on the amount of coastal marsh.

Coastal wetlands also provide food, shelter, and nesting and resting areas for other wildlife. Migrating waterfowl (ducks and geese), shorebirds, and wading birds, for example, either overwinter in southern coastal wetlands or rest and feed here before traveling farther south. Many birds are year-round residents, including American egrets, snowy egrets, great blue herons, clapper rails, seaside sparrows, and marsh wrens. These birds nest in or adjacent to coastal wetlands and feed in the marshes or mangroves. Other animals also depend on coastal marshes, including muskrats, nutrias, swamp rabbits, American alligators, fiddler crabs, marsh periwinkles, and salt-marsh snails. Fish and wildlife habitat values of coastal wetlands have been recognized for a long time.

Historical uses of coastal wetlands are fishing, hunting, shellfish harvesting, and grazing by cattle or horses. In the 1920s, Louisiana's coastal marshes alone generated over 8 million dollars in annual revenues for the state's economy. Other values now recognized include: (1) flood and storm damage protection by temporarily storing floodwaters and by buffering dry land from storm wave action; (2) water quality maintenance by removing sediments, nutrients, and other materials from flooding waters; (3) shoreline stabilization by slowing current velocity and minimizing erosive action of water; (4) nonconsumptive recreation (e.g., wilderness experiences, nature photography, and bird watching); (5) environmental education; and (6) aesthetics.

One might reasonably think that with all these values coastal wetlands have been regarded as vital natural resources—aquatic treasures—for a long time, but this is, unfortunately, not so. Despite some knowledge of the natural values, coastal wetlands have been greatly abused in the past. Once considered as wastelands that bred mosquitoes, these wetlands were highly regarded as potential sites for development. Their proximity to deep water and to coastal beaches placed coastal wetlands in a precarious position. As the U.S. population grew after World War II, much of the growth was in the coastal zone. Many acres of coastal wetlands were dredged and filled for housing developments and commercial offices and businesses in coastal communities. Marinas, expanded port facilities, and maintenance of navigable waters involved dredging and/or disposal of dredged material in these wetlands. While this was occurring, scientists studying coastal wetlands found that they were productive areas essential to our coastal fisheries and other aquatic organisms. Coastal wetlands quickly became viewed as critical natural resources, worthy of special protection. Massachusetts was the first state to pass a law to protect these wetlands in 1963. Presently, all coastal states except Texas have laws regulating uses of coastal wetlands. The federal government also strengthened its role in conserving these areas through expanded regulation of the Rivers and Harbors Act of 1899 and by developing new regulations pursuant to the Federal Water Pollution and Control Act of 1972 (later amended as the Clean Water act of 1977). Today, coastal wetlands across the country are receiving much better protection, and alternative uses are strictly controlled, for the most part, in most areas.

All of these laws define wetlands, in part, on the occurrence of plants adapted for life in wet areas. The state laws list numerous plants that are characteristic of tidal marshes. Plant identification is therefore an important step toward identifying coastal wetlands. For more than twenty years, I have worked in the field of wetland mapping and protection, and, to my continued surprise, a thorough, easy-to-use field guide for identifying common southeastern coastal wetland plants has not yet been developed. The purpose of this book is to fill this void and to introduce readers to coastal wetland ecology. It is also intended to identify specific places where tidal wetlands can be visited. The focus of the book, however, remains on plant identification. This guide is designed primarily for the nontechnical person interested in learning to recognize common tidal wetland plants. It will be useful for representatives of federal and state environmental agencies, members of local environmental and planning commissions, environmental consultants, students in botany and environmental sciences, interpretation specialists at parks and nature centers, and other individuals interested in coastal wetlands. This book is intended for primary use in coastal wetlands in the southeastern United States, from Virginia through Texas. Although the focus is on

coastal wetlands, keep in mind that most of the plants illustrated for tidal fresh marshes and swamps are also characteristic of other freshwater wetlands in the region. The guidebook should, therefore, be useful along the Coastal Plain, from Maryland and Delaware to Florida and west to Texas and adjacent Mexico, despite omissions of many strictly tropical species.

This book is arranged in four major sections: (1) Coastal Wetland Ecology: A General Overview, (2) Identification of Coastal Wetland and Aquatic Plants, (3) Wetland Plant Descriptions and Illustrations, and (4) Distribution of Coastal Wetlands in the Southeast. In addition, a list of references used in preparing this book and a glossary of technical terms are provided.

Coastal Wetland Ecology: A General Overview

Coastal wetlands are low-lying areas periodically flooded by tidal waters for varying lengths of time. They develop in sheltered areas behind barrier islands, along the shores of coastal embayments and tidal rivers, or in open water in low-energy, exposed areas. (See Plate 1.) These wetlands may be flooded by salt water of varying concentrations or by fresh water subject to tide-induced fluctuations. They generally include tidal mud or sand flats, salt marshes, salt barrens or flats, mangrove swamps, brackish marshes, oligohaline (intermediate) marshes, and tidal freshwater marshes and swamps. Other coastal wetlands are represented by intertidal beaches, rocky shores, and oyster reefs. Aquatic beds in permanently flooded (subtidal) areas are often associated with intertidal wetlands, either in pools or ponds within the marshes, in adjacent shallow water, or in man-made impoundments.

Factors Affecting Coastal Wetlands and Their Vegetation

Flooding by tidal water is the common denominator of all coastal (tidal) wetlands. It is the driving force that creates, shapes, and maintains these habitats and makes them different from their nontidal counterparts—the inland (nontidal) wetlands. Tides are the result of the gravitational forces of the moon and the sun exerted upon the earth and modified locally by weather conditions. A tidal cycle consists of one high tide and one low tide. If only one tidal cycle occurs during a twenty-four-hour period, then the tides are called *diurnal*. When two tidal cycles take place each day, the tides are *semidiurnal*. Depending on the position of the moon relative to the sun, other tides occur at two-week intervals. *Spring tides* are the highest astronomic tides. They occur on new and full moons and produce higher high tides and lower low tides, resulting in a greater tidal range than normal. *Neap tides* are moderated tides, producing the lowest tidal range. They take place around the moon's first and third quarters and follow the spring tides by about a week. Local climatic conditions, especially winds and storms (e.g., hurricanes), influence the actual height, duration, and frequency of tidal flooding. Wind effects are most pronounced in areas of low tidal range. (See Table 1.)

The behavior of the tides in the Southeast and elsewhere is governed by the size and shape of the estuary and offshore bottom characteristics. Along the Atlantic coast and the eastern portion of the Gulf of Mexico the tides are semidiurnal, whereas along the northern Gulf coast diurnal tides occur. Tidal ranges are generally higher on the Atlantic than along the Gulf, so the effect of wind is more significant in the Gulf. Higher sea levels in the late summer and early fall

Table 1. Examples of tidal ranges for the Southeast: Mean and spring tidal ranges for South Atlantic locations and mean diurnal tides for Gulf coast sites.

Location	Tidal Range (ft) Mean	Tidal Range (ft) Spring	Location	Tidal Range (ft) Diurnal
Potomac River			Florida Bay (eastern part)	<0.5
Colonial Beach, VA	1.7	1.9	Charlotte Harbor	
Alexandria, VA	2.8	3.2	Punta Gorda, FL	1.9
Mattaponi River			Tampa Bay, FL	2.3
Walkerton, VA	3.9	4.5	Cedar Key, FL	3.5
Albemarle and Pamlico Sounds, NC	<0.5	—	Apalachicola Bay	
Hatteras Inlet, NC	2.0	2.4	Cut Point, FL	2.2
Hatteras, NC (oceanside)	3.4	4.1	West Pass, FL	1.4
Cape Fear River			Perdido Bay, AL	<0.5
Wilmington, NC	4.1	4.8	Mobile Bay, AL	1.4
Little River, SC	5.0	5.9	Mississippi Sound, MS	1.5
Cooper River			Mississippi River	
Dean Hall, SC	4.1	4.8	Entrance	1.2
Folly Beach, SC (oceanside)	5.2	6.1	New Orleans, LA	0.8[a]
May River			Barataria Bay, LA	1.0
Bluffton, SC	8.1	9.5	Atchafalaya Bay, LA	1.9
Savannah River			Sabine Pass, TX	2.5
Entrance	6.9	8.0	Galveston, Bay, TX	1.0
Savannah, GA	7.4	8.6	Baffin Bay, TX	0.3
Satilla River				
Bailey Cut, GA	6.9	8.1		
Burnt Fork, GA	3.2	3.7		
Daytona Beach, FL (oceanside)	4.1	4.9		
Miami Beach, FL	2.5	3.0		

[a] At low river stage; no tide at high river stage.

combined with southerly winds cause increased flooding of many Gulf coast marshes at this time.

Within coastal wetlands, flooding frequency is related to elevation. The lowest portions of tidal flats may be almost continuously flooded and exposed to air only during extreme low tides. The majority of coastal wetlands, however, are subjected to alternate flooding and exposure to air for variable periods. Two general zones are recognized, based on the frequency of flooding: (1) the *regularly flooded zone* and (2) the *irregularly flooded zone*. The former zone is flooded at least once daily by the tides, whereas the latter zone is flooded less often and is generally exposed to air for long periods. (See Figure 1.) Even when not flooded, the soils of coastal wetlands remain saturated at or near the surface. This hydrology usually creates anaerobic or oxygen-deficient conditions, which greatly affect the types of plants that can become established. Oxygen depletion of the soil results in reduction or mobilization of chemical elements toxic to most plants. Coastal wetland soils are particularly noted for their hydrogen sulfide accumulations, which produce a "rotten egg" odor. Plants adapted for life in flooded or saturated soils—*hydrophytes*—usually form extensive communities under these conditions.

The frequency and duration of flooding affect plant composition in wetlands as well as the growth form of some species, notably smooth cordgrass (*Spartina alterniflora*) and black needlerush (*Juncus roemerianus*).

Other factors besides tidal hydrology and accompanying anaerobic soil conditions are operating to influence wetland vegetation patterns. They include salinity, substrate (mud, sand, muck, or peat), temperature, hurricanes, biological competition, and human activities, such as wildlife management practices, ditching, other hydrologic alterations, pollution, and filling. Of these, salinity is perhaps the most important natural factor affecting plant growth, though the human impact on current vegetation patterns is enormous.

Because coastal wetlands occupy positions between the open ocean and nontidal freshwater bodies, they largely exist in a zone of transition or flux where sea water intermixes with fresh water. This area of mixing is called the *estuary*. An estuary includes both deepwater areas and contiguous intertidal wetlands. At the seaward end of the estuary, water salinity approaches sea strength (35 parts per thousand). As one moves farther upstream in coastal rivers, water salinity is diluted by increasing volumes of fresh water. This layer of salt water moves back and forth within the estuary as a wedge with the heaviest salt water at the bottom and a lens of fresh water at the surface. During periods of low freshwater discharge from rivers (e.g., late summer), the salt wedge penetrates to its farthest upstream point. By contrast, heavy river discharge such as spring runoff forces the salt wedge far downstream. In large coastal rivers with high freshwater discharge, eventually a point is reached where the water is strictly fresh, with no trace of ocean salts, yet the water levels are still subject to rises and falls with the changing tides. Here, tidal waters moving upstream on the rising tide form a barrier that impedes much of the downstream flow of fresh water. This damming effect forces water levels to rise with the flooding tide. On the ebb or falling tide, the barrier is gradu-

Figure 1. Coastal wetlands can be divided into two broad zones based on the frequency of tidal flooding: (1) regularly flooded zone (low marsh), and (2) irregularly flooded zone (high marsh). The highest lunar-driven tides, called "spring tides," occur during full and new moons; coastal storms can generate even higher tides that may inundate low-lying uplands.

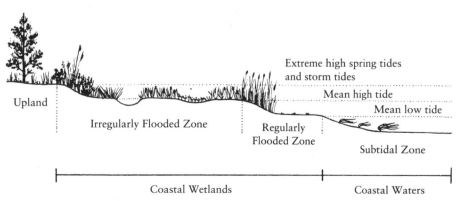

ally removed, allowing the fresh water to flow freely downstream. Upstream beyond this area, the river is not influenced by the tides, and this begins the nontidal reaches of the river. (See Figure 2.)

Plants must absorb water from the soil to grow. Since the soil water in salt and brackish wetlands is saline, plants colonizing these habitats must effectively deal with salt stress. There are several ways of doing this: (1) reduce salt uptake by roots (salt exclusion); (2) develop salt-secreting glands to get rid of excess salt; (3) develop salt-concentrating organs (e.g., fleshy leaves) and periodically shed them; (4) increase succulence (fleshiness) to keep internal salt concentrations at acceptable levels; (5) cover leaf surfaces with waxy flakes to make them saltwater-proof; (6) reduce leaf surfaces to minimize exposure to salt and evapotranspiration; and (7) isolate salt within certain internal organs. Plants may follow more than one of these and other approaches to solving the salt stress problem. (See Plate 2.) The presence of salt glands may be the most common of these adaptations—look for salt crystals on the leaves of smooth cordgrass (*Spartina alterniflora*). Some plants avoid serious salt stress by occupying only the higher portions of salt marshes where infrequent tidal flooding and significant freshwater runoff from the adjacent upland create a fresher environment. All plants growing in salt and brackish marshes must be adapted to or tolerant of saline water. These plants are often referred to as *halophytes* or salt-loving plants. Only a few plants actually require salt water for growth. Among these *obligate halophytes* are glassworts (*Salicornia* spp.) and saltwort (*Batis maritima*). Most halophytes are *facultative halophytes* that do not require salt water but tolerate it and do well in these habitats due to lack of competition from other wetland plant spe-

cies. The increase in plant diversity with decreasing salinity in coastal wetlands as one moves upstream in tidal rivers or from the regularly flooded salt marsh to the upper limits of the irregularly flooded zone provides ample evidence to support this view.

Wind action in areas of low tidal amplitude has significant effects not only on tidal levels but also on salinity. For example, strong northerly winds in the Gulf region in winter lower salinity by blowing much salt water out of the bays and keeping it out, thereby allowing freshwater inflow from rivers to further dilute bay salinities. In some places, such as along the Texas coast, these strong winds leave vast areas of flats exposed to air for significant periods.

In the Everglades region of South Florida, salinity patterns are different than elsewhere in the Southeast, due to distinct wet and dry seasons. From December through May, there is little rain, resulting in low river flows and increased salinity and salt intrusion in coastal rivers. Late May brings the rainy season, and by June many rivers are fresh and remain so until November. Hurricane-generated tides may bring salt water up to 15 miles inland in the Everglades. Periodic hurricanes, therefore, have a significant impact on coastal wetland vegetation.

Rising sea level due to global warming is increasing both tidal flooding duration and salinity. This greatly affects coastal vegetation patterns and poses a serious risk for coastal wetlands. Wherever the build-up of marsh sediments cannot keep pace with the rising sea level, marshes will become tidal flats and eventually open-water habitats. Salt water penetration further upstream can cause the death of intolerant tidal fresh marsh vegetation. Depending on the rate of colonization by halophytic species and the nature of the substrate, these marshes may change to brackish marshes or be eroded

SUMMER

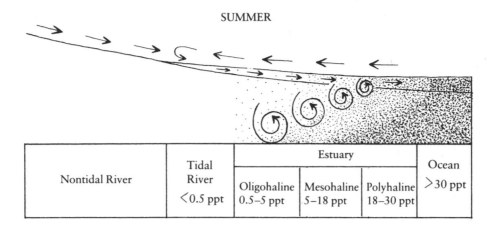

Nontidal River	Tidal River <0.5 ppt	Estuary			Ocean >30 ppt
		Oligohaline 0.5–5 ppt	Mesohaline 5–18 ppt	Polyhaline 18–30 ppt	

SPRING

Nontidal River	Tidal River	Estuary	Ocean

Mixing zone

Salt water

Fresh water

→ Downstream current (ebbing tide and nontidal river flow)

← Upstream current (rising tide)

Figure 2. Generalized salinity and current patterns in coastal rivers. Note: (1) The upstream limit of the estuary is defined by maximum penetration of measurable sea water; (2) the position of the mixing zone changes seasonally; (3) salinities throughout the estuary gradually decrease upstream; and (4) two-directional currents exist in the ocean, estuary, and tidal river due to the tides, whereas the flow of the nontidal river is one-directional (downstream). Heavy spring discharges of fresh water may temporarily eliminate tidal fluctuations from normally tidal portions of coastal rivers.

and become either tidal flats or open-water habitat. Recent losses of Louisiana's coastal marshes illustrate the severity of the problem, with an estimated 25,000 acres lost each year. The causes of these losses are, however, not solely attributed to rising sea level and coastal subsidence. Human activities, especially canal and levee construction, oil and gas extraction, and major river diversion, have exacerbated the problem. Along the Gulf–Atlantic Coastal Plain, low-lying forests border many of the coastal wetlands. With increased tidal flooding by more saline waters, these forests are being replaced by salt and brackish marshes migrating landward. (See Plate 3.) The presence of these unprotected shorelines should permit the landward advance of coastal wetlands and help lessen their overall loss due to rising sea levels.

Southeastern Coastal Wetland Types

Under the variety of flooding, salinity, temperature, and other conditions, seven major types of coastal wetlands have formed in the Southeast: (1) mangrove swamps, (2) salt marshes, (3) salt barrens and flats, (4) brackish marshes, (5) tidal fresh marshes, (6) tidal swamps, and (7) tidal flats. (See Table 2 and Figure 3.) Shrub wetlands, besides the mangrove swamps, may be locally important but are much less abundant than other types. Moreover, they have been traditionally included in descriptions of various coastal marshes and swamps. Each major coastal wetland type is generally described below. Aquatic beds are also discussed because they are often found within the coastal wetlands or in adjacent waters and are similarly valuable estuarine habitats.

Table 2. Wetland types referred to in this book, with their corresponding technical classifications according to the U.S. Fish and Wildlife Service. This classification system was developed primarily for mapping the nation's wetlands.

Coastal Wetland Type Used in This Book	U.S. Fish and Wildlife Service Classification			
	System(s)	Subsystem	Class	Water Regime(s)
Tidal Flat	Marine and Estuarine	Intertidal	Unconsolidated Shore	Regularly Flooded
	Riverine	Tidal	Unconsolidated Shore	Regularly Flooded
Mangrove Swamp	Estuarine	Intertidal	Scrub-Shrub Wetland and Forested Wetland	Regularly Flooded and Irregularly Flooded
Salt Marsh and Brackish Marsh	Estuarine	Intertidal	Emergent Wetland (includes Scrub-Shrub Wetland)	Regularly Flooded and Irregularly Flooded
Salt Barren or Flat	Estuarine	Intertidal	Unconsolidated Shore or Emergent Wetland	Irregularly Flooded
Tidal Fresh Marsh	Riverine Palustrine	Tidal	Emergent Wetland Emergent Wetland (includes Scrub-Shrub Wetland)	Regularly Flooded Seasonally Flooded-Tidal
Tidal Swamp	Palustrine		Forested Wetland	Seasonally Flooded-Tidal and Temporarily Flooded-Tidal

Note: Coastal aquatic beds are mainly subtidal and classified as deepwater habitats.

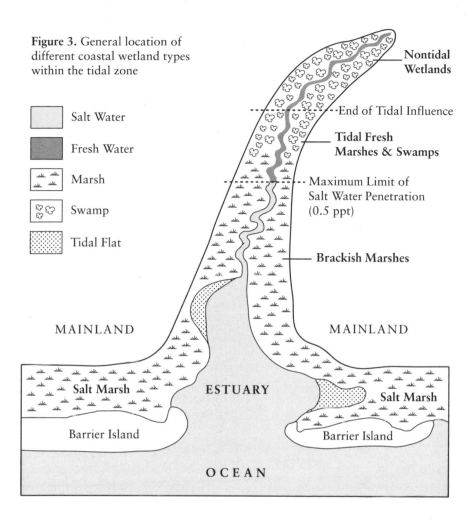

Figure 3. General location of different coastal wetland types within the tidal zone

Salt Water

Fresh Water

Marsh

Swamp

Tidal Flat

Nontidal Wetlands

End of Tidal Influence

Tidal Fresh Marshes & Swamps

Maximum Limit of Salt Water Penetration (0.5 ppt)

Brackish Marshes

MAINLAND

MAINLAND

Salt Marsh

ESTUARY

Salt Marsh

Barrier Island

Barrier Island

OCEAN

Mangrove Swamps

Mangrove swamps are a dominant coastal wetland type in the southernmost United States and elsewhere in the world's tropics. They dominate about three-quarters of the world's coastline between 25 degrees north and 25 degrees south latitude and extend somewhat farther north and south in certain areas. In Florida, mangrove swamps are common south of Cape Canaveral on the Atlantic coast and south of Tarpon Springs on the Gulf coast. (See Plate 4.) They are best developed along the southwestern coast from Cape Sable to Everglades City where they form the Ten Thou-

sand Islands. Salt marshes and sea grass beds are also abundant in South Florida.

Three mangrove species characterize Florida's mangrove swamps: red mangrove (*Rhizophora mangle*), black mangrove (*Avicennia germinans*), and white mangrove (*Laguncularia racemosa*). Black mangrove is the most northerly ranging of these species, growing north to about 30 degrees north latitude (St. Augustine–Jacksonville Beach) on the Atlantic coast and growing in shrubby patches along the northern Gulf coast. The other mangroves occur only in peninsular Florida from Daytona Beach (Ponce de Leon Inlet) on the Atlantic to Cedar Key on the Gulf.

Red mangrove with its conspicuous prop roots typically occupies the regularly flooded zone along bay shores and tidal rivers. (See Plate 5.) Black mangrove usually dominates the irregularly flooded swamp interior where white mangrove also occurs. Both white and black mangroves can be found in overwash islands and beaches where red mangrove usually prevails. At the upper edges of mangrove swamps, another woody species—buttonwood (*Conocarpus erectus*)—forms the swamp border. Brazilian pepper (*Schinus terebinthifolius*) and Australian pine (*Casuarina equisetifolia*), invasive exotics, may also occupy the border areas. Other species present in mangrove swamps include leather ferns (*Acrostichum aureum* and *A. danaeifolium*), perennial glasswort (*Salicornia virginica*), saltwort (*Batis maritima*), bay marigold (*Borrichia arborescens*), sea ox-eye (*Borrichia frutescens*), silverhead or marsh samphire (*Philoxerus vermicularis*), sea blite (*Suaeda linearis*), sea purslane (*Sesuvium portulacastrum*), spider lily (*Hymenocallis latifolia*), high-tide bush (*Iva frutescens*), cabbage palm (*Sabal palmetto*), deer pea (*Vigna luteola*), gray nicker (*Caesalpinia bonduc*), coin-vine (*Dalbergia ecastophyllum*), and rubber vine (*Rhabdadenia biflora*). Salt marshes are often intermixed with the mangroves and are typically found landward of these swamps. Higher elevations called *hammocks* also occur within the mangrove swamps. These areas are dominated by tropical hardwood trees.

Scientists have described six different types of mangrove swamps in Florida based mainly on landscape position, which creates different hydrologic and other environmental conditions: (1) overwash forests, (2) fringing forests, (3) riverine swamps, (4) basin forests, (5) hammock forests, and (6) dwarf forests. Overwash forests occur on islands frequently overwashed by tides.

Red mangroves about 25 feet tall usually predominate in these sites. Fringing forests of red mangrove (about 30 feet high) form narrow bands, usually above mean high water, along waterways and water bodies. Riverine mangrove swamps are essentially regularly flooded floodplain forests typically dominated by red mangrove (up to 65 feet tall). Basin mangrove forests develop in depressions where the frequency and duration of flooding depend on elevation: Red mangrove occupies regularly flooded basins, and the black and white mangroves dominate the irregularly flooded basins. Basin mangroves grow to about 50 feet in height. Hammock forests occur on slightly elevated sites where all three mangrove species may occur. Here they grow less than 15 feet tall. When mangroves colonize limestone marl, they grow in a stunted form less than 5 feet tall. These areas are called dwarf forests. (See Plate 6.)

Along the northern Gulf coast, black mangrove is the only mangrove present, forming dense, shrubby thickets in salt marshes. Associated plants include saltwort and glassworts (*Salicornia* spp.), among other typical salt marsh species.

Salt Marshes

Salt marshes are intertidal areas colonized by grasses and other salt-tolerant plants (halophytes). These marshes are the predominant coastal wetland type through the Southeast, especially in South Carolina, Georgia, and Louisiana. They typically occur behind barrier islands (sea islands) and at the mouths of coastal rivers in high-salinity waters. The dominant plants are emergent (herbaceous) species, although shrubs may also be locally abundant. (See Plate 7.)

Two vegetation zones are evident within these marshes: *low marsh* and *high marsh*. The low marsh is the regularly flooded zone subject to daily tidal flooding, whereas the high marsh is flooded less often. The difference in flooding affects the plant composition of these zones.

The low marsh often consists of a single species—smooth cordgrass (*Spartina alterniflora*), which grows to a height of 6 feet or more (called the tall form) along creek banks but usually ranges between 3 and 6 feet tall in the low marsh interior. (See Plate 8.) Low marsh is the predominant coastal marsh in Georgia, Louisiana, and South Carolina where it forms a grassy belt up to several miles wide between the barrier islands and the mainland. Other plants that may be found in the low marsh include perennial glasswort (*Salicornia virginica*) and sea lavenders (*Limonium carolinianum* and/or *L. nashii*).

The high marsh is floristically more diverse than the low marsh. Several species may dominate, including the short form of smooth cordgrass (less than 1½ feet tall), salt grass (*Distichlis spicata*), salt meadow cordgrass or wiregrass (*Spartina patens*), black needlerush (*Juncus roemerianus*), sea ox-eye, coastal dropseed (*Sporobolus virginicus*), salt marsh bulrush (*Scirpus robustus*), big cordgrass (*Spartina cynosuroides*), glassworts, saltwort, key grass (*Monanthochloe littoralis*, along the Florida and Gulf coasts), and Gulf cordgrass (*Spartina spartinae*, along the Gulf coast).

The latter four species are typical of salt flats within the high marsh. Wetter depressions in the high marsh may be colonized by salt grass, smooth cordgrass, salt marsh bulrush, salt marsh fleabane (*Pluchea purpurascens*), coastal water-hyssop (*Bacopa monnieri*), and spike-rushes (*Eleocharis* spp.). Sandy sites may have sea purslane, salt marsh sand spurrey (*Spergularia marina*), salt meadow cordgrass, key grass, coastal dropseed, sea lavender, Gulf cordgrass, marsh finger grass (*Eustachys glauca*), annual marsh pink (*Sabatia stellaris*), fragrant galingale or flatsedge (*Cyperus*

odoratus), Nuttall's cyperus (*Cyperus filicinus*), white spike-rush (*Eleocharis albida*), knot-grass (*Paspalum distichum*), silverhead (or marsh samphire), seaside heliotrope (*Heliotropium curassavicum*), marsh pennywort (*Hydrocotyle umbellata*), and salt marsh fimbristylis (*Fimbristylis castanea*). Shrubs besides sea ox-eye may be locally common in the high marsh. They include high-tide bush, groundsel bush or sea myrtle (*Baccharis halimifolia*), and black mangrove (along the Florida and Gulf coasts). The former two shrubs, however, are more common along the upper edges of the marshes where they occur with other woody plants, such as saltwater false willow (*Baccharis angustifolia*), wax myrtle (*Myrica cerifera*), cabbage palm, eastern red cedar (*Juniperus virginiana*), southern red cedar (*Juniperus silicicola*), yaupon (*Ilex vomitoria*), dahoon (*Ilex cassine*), and live oak (*Quercus virginiana*).

The upper marsh edge typically exhibits the greatest plant diversity within the salt marsh, due to less salt stress. Characteristic herbaceous plants found here include salt meadow cordgrass, black needlerush, salt marsh fimbristylis, perennial salt marsh aster (*Aster tenuifolius*), annual marsh pink, big cordgrass, common reed (*Phragmites australis*, formerly *P. communis*), Gulf cordgrass, climbing milkweed (*Cynanchum angustifolium*), seaside goldenrod (*Solidago sempervirens*), sand cordgrass (*Spartina bakerii*), slender-leaved goldenrod (*Euthamia galetorum*, formerly *Solidago tenuifolia*), deer pea, bloodleaf (*Iresine rhizomatosa*), common frog-fruit (*Phyla nodiflora*), switchgrass (*Panicum virgatum*), and broomsedges (*Andropogon* spp.). In South Florida, border species also may include coastal leather fern (*Acrostichum aureum*), sea grape (*Coccoloba uvifera*), and buttonwood. Along the upland where freshwater runoff or groundwater discharge is significant (e.g., seepage areas), brackish and fresh species may be found. Some of the more common plants in these situations are Olney's three-square (*Scirpus americanus*, formerly *S. olneyi*), common three-square (*Scirpus pungens*, formerly *S. americanus*), salt marsh fleabane, narrow-leaved cattail (*Typha angustifolia*), rose mallow (*Hibiscus moscheutos*), seashore mallow (*Kosteletzkya virginica*), salt marsh loosestrife (*Lythrum lineare*), spikerushes, and poison ivy (*Toxicodendron radicans*).

Along the northeastern Gulf coast (i.e., Alabama, Mississippi, and Florida), salt marshes are dominated by high marshes, in contrast to the South Atlantic coast where low marsh predominates. Vast expanses of black needlerush characterize these marshes, much like smooth cordgrass does along the Atlantic coast. On the latter coast, black needlerush is often confined to a band of varying widths near the upper edges of the salt marsh, but it becomes more abundant in the brackish marshes. Plant diversity within the needlerush marshes is extremely low, with scattered individuals of a few other salt marsh species present.

Many salt marsh plants have a wide distribution in the United States, whereas others have more restricted ranges. Some prominent species that occur both in the Northeast and Southeast (Atlantic and Gulf coasts) include smooth cordgrass, salt meadow cordgrass, salt grass, common reed, marsh fleabane, seaside goldenrod, perennial salt marsh aster, annual marsh pink, marsh orach (*Atriplex patula*), perennial glasswort, rose mallow, high-tide bush, and groundsel bush. Some strictly southern species are sea ox-eye, saltwort, sandpaper vervain (*Verbena scabra*), coastal dropseed, Gulf cordgrass, sand cordgrass, key grass, sea purslane, common frog-fruit, deer pea, climbing milkweed, saltwater false willow, and Christmas-berry (*Lycium carolinianum*).

Salt Barrens and Flats

Located within the high salt marsh, the more seaward brackish marshes, and mangrove swamps are largely nonvegetated zones called salt barrens, salt flats, or salinas. For the most part, these areas are devoid of plants, covered by thin mats of blue-green algae, and/or sparsely vegetated by macrophytes (salt flats) due to extremely high soil salinity (in excess of 100 parts per thousand in some places). This salinity results from a combination of factors, including infrequent tidal flooding, high evaporation, and, in some areas, low seasonal rainfall. Salt barrens and flats seem to develop above a critical point of frequent flooding and below the point where effective leaching by rainfall occurs. The size and shape of the barrens may change annually due to changing weather patterns, especially rainfall and tidal flooding. Barrens may increase in size during periods of lower rainfall and less frequent flooding but may shrink in size with higher rainfall and more frequent flooding. (See Plate 9.)

Along the edges of the barrens, salt flats vegetated by only the most salt-tolerant plants are found. Among the typical species are glassworts (*Salicornia virginica*, *S. europaea*, and *S. bigelovii*), saltwort, salt grass, key grass, coastal dropseed, sea ox-eye, stunted black needlerush, smooth cordgrass (short form), sea purslane, Christmas-berry or Carolina wolfberry (*Lycium carolinanum*), seaside gerardia (*Agalinis maritima*), purple gerardia (*A. purpurea*), annual marsh pink, perennial marsh pink (*Sabatia dodecandra*), Gulf cordgrass, sea blite, and seaside heliotrope.

Salt flats are quite extensive along the northern coast of Florida Bay in Everglades National Park. Saltwort, perennial glasswort, and sea purslane are the chief dominants. Salt grass and coastal dropseed occur as codominants with these species or form large colonies themselves. Other herbaceous plants found in these areas are key grass, sea blite, and silverhead. Buttonwood and mangroves usually border these flats.

Brackish Marshes

Brackish marshes develop upstream of salt marshes where significant amounts of fresh water dilute seawater to create moderately to slightly salty environments. Average salinities in this region range from moderately high (18 parts per thousand) to essentially fresh (0.5 ppt). Consequently, plant composition is extremely varied. Brackish marshes are found along coastal rivers upstream of the salt marshes or near the mouths of coastal rivers with heavy freshwater discharge that empty into bays and sounds with low tidal ranges. Due to diurnal tides of low amplitude (averaging about 1 foot), brackish marshes predominate along the northeastern Gulf coast, whereas salt marshes occupy similar positions along the South Atlantic coast where semidiurnal tides of moderate amplitude (about 6 to 8 feet) prevail.

The more seaward brackish marshes resemble the upper high marsh zone of the salt marsh. Dominant species include black needlerush, salt grass, smooth cordgrass, glassworts, salt meadow cordgrass, sea oxeye, and switchgrass. Black needlerush forms monotypic stands (see Plate 10), with other dominants usually occurring in patches of varying size, giving the marshes a mosaic appearance. Smooth cordgrass can still be found dominating the regularly flooded low marsh, which typically forms a distinctive band along tidal rivers and creeks. Where wave action is considerable, natural levees may be colonized by the other species listed above, plus high-tide bush, groundsel bush, and big cordgrass. Other common associates in the needlerush marshes are typical salt marsh plants, such as salt marsh bulrush, saltwort, salt marsh

aster, salt marsh fimbristylis, seaside goldenrod, salt marsh fleabane, Gulf cordgrass, and climbing milkweed. Late-flowering thoroughwort or boneset (*Eupatorium serotinum*), a medium-height, white-flowered plant, is conspicuous among the rushes, although it is not particularly abundant.

The marsh border may be represented by high-tide bush, groundsel bush, wax myrtle, big cordgrass, and common reed. Halberd-leaved morning glory (*Ipomoea sagittata*) with its pinkish purple funnel-shaped flowers often can be seen climbing the shrubs. Along the border where freshwater influence is significant, sawgrass (*Cladium jamaicense*) and switchgrass become dominant as the coastal marsh grades into adjacent pine flatwoods. Yaupon, loblolly pine (*Pinus taeda*), and cabbage palm may occur in these locales.

Black needlerush exhibits different growth forms, as does its salt marsh counterpart, smooth cordgrass. Three growth forms have been reported: (1) tall form (5–6 feet tall), (2) short or dwarf form (about 1 foot tall), and (3) intermediate form (about 3 feet tall). Height differences may be related to flooding frequency, substrate, salinity, and nutrient availability. Tall plants occur on more frequently flooded peaty soils with low salinity. Dwarf plants grow on sandy soils with high salinity (near salt barrens). The intermediate form grows on sandy clays of moderate salinity.

Upstream of the needlerush marshes, other brackish species predominate. Salt meadow cordgrass, salt grass, salt marsh bulrush, perennial salt marsh aster, spike-rushes (e.g., *Eleocharis parvula, E. cellulosa*, and *E. tuberculosa*), Olney's three-square, knotgrasses (*Paspalum distichum* and *P. vaginatum*), torpedo grass (*Panicum repens*), and big cordgrass are among the abundant species. Salt meadow cordgrass is the predominant brackish marsh plant in

Louisiana. (See Plate 11.) Also common in these marshes are seashore mallow, rose mallow, salt marsh loosestrife, common reed, and eastern lilaeopsis (*Lilaeopsis chinensis*). Typical salt marsh species, such as glasswort, are absent, yet smooth cordgrass may still be found along creeks and needle-rush may still be common.

As freshwater influence increases, plants with freshwater affinities intermix with brackish species to create highly diverse marshes. These slightly brackish marshes are called *oligohaline marshes* (due to their low salinities) or *intermediate marshes* (due to their position between brackish and tidal fresh marshes). Black needlerush may be present, but it is generally replaced as a dominant by salt meadow cordgrass, big cordgrass, common reed, switchgrass, bull-tongue or lance-leaved arrowhead and coastal arrowhead (*Sagittaria lancifolia* and *S. falcata*), cattails (*Typha angustifolia* and others), or by a mixture of these and other plants, including spike-rushes, saw-grass, salt marsh bulrush, water hemp (*Amaranthus cannabinus* and *A. australis*), rose mallow, Olney's three-square, common three-square, deer pea, soft-stemmed bulrush (*Scirpus validus*), California bulrush (*S. californicus*), bearded sprangletop (*Leptochloa fascicularis*), Gulf cordgrass, coastal water-hyssop, alligatorweed (*Alternanthera philoxeroides*), fragrant flatsedge, Walter millet (*Echinochloa walteri*), southern blue flag (*Iris virginica*), smartweeds (*Polygonum* spp.), and pickerelweed (*Pontederia cordata*). Smooth cordgrass may occur along the water's edge, but it is generally supplanted by pickerelweed, wild rice (*Zizania aquatica*), arrow arum (*Peltandra virginica*), water hemp, soft-stemmed bulrush, cattails, alligatorweed, and coastal water-hyssop. Other notable species in these marshes are spider lilies (*Hymenocallis crassifolia* and *H. occidentalis*), southern swamp lily (*Crinum americanum*),

white-top sedge (*Dichromena colorata*), water parsnip (*Sium suave*), and giant cut-grass (*Zizaniopsis miliacea*). These marshes grade imperceptibly into tidal fresh marshes.

Tidal Fresh Marshes

Eventually a point is reached in tidal rivers where salt water does not penetrate due to freshwater discharge throughout the year. Here, water levels rise and fall with the tides, but waters are strictly fresh (less than 0.5 parts per thousand of sea salts). Due to the lack of salt water, many additional wetland plants can become established. These tidal fresh marshes, as they are called, are among the most diverse wetland plant communities in the continental United States. Smooth cordgrass and other salt marsh plants are typically absent (e.g., salt marsh bulrush, salt grass, and widgeon grass). Giant cutgrass, maidencane (*Panicum hemitomom*), and other freshwater species predominate, and others that occurred in brackish marshes are still common or locally dominant, including lance-leaved arrowhead (*S. falcata*), common reed, salt meadow cordgrass, narrow-leaved cattail, big cordgrass, deer pea, sand cordgrass, alligatorweed, sawgrass, arrow arum, pick-erelweed, common three-square, smart-weeds, soft-stemmed bulrush, Walter millet, and wild rice. Other characteristic tidal fresh marsh species are other cattails (*Typha* spp.), jointed spike-rush (*Eleocharis equisetoides*), wood reed (*Cinna arundinacea*), umbrella sedges (*Cyperus* spp.), sedges (*Carex* spp.), swamp dock (*Rumex verticillatus*), bur marigold (*Bidens laevis*), beggar-ticks (*Bidens* spp.), rice cutgrass (*Leersia oryzoides*), water primroses (*Ludwigia* spp.), Asiatic dayflower (*Murdannia keisak*), water parsnip, water hemlock (*Cicuta maculata*), tearthumbs (*Polygonum arifolium* and *P. sagittatum*), American cupscale (*Sacciolepis striata*), jewelweed (*Impatiens capensis*), false nettle (*Boehmeria cylindrica*), marsh fern (*Thelypteris thelypteroides*), royal fern (*Osmunda regalis*), climbing hempweed (*Mikania scandens*), soft rush (*Juncus effusus*), sweet flag (*Acorus calamus*), marsh pennywort, mock bishop-weed (*Ptilimnium capillaceum*), asters (*Aster* spp.), butterweed (*Senecio glabellus*), cardinal flower (*Lobelia cardinalis*), and marsh eryngo (*Eryngium aquaticum*). Shrubs and scattered trees may occur within these marshes. They include swamp rose (*Rosa palustris*), buttonbush (*Cephalanthus occidentalis*), wax myrtle, grounsel bush, false indigo (*Amorpha fruticosa*), red maple (*Acer rubrum*), water locust (*Gleditsia aquatica*), smooth alder (*Alnus serrulata*), swamp or Carolina willow (*Salix caroliniana*), common elderberry (*Sambucus canadensis*), bald cypress (*Taxodium distichum*), and southern arrowwood (*Viburnum dentatum*). Along the water's edge, other species may be observed: spatterdock (*Nuphar luteum*), golden club (*Orontium aquaticum*), pickerelweed, arrowheads, arrow arum, southern swamp lily, cattails, bulrushes (*Scirpus* spp.), alligatorweed, maidencane, giant cutgrass, wild rice, American frog-bit (*Limnobium spongia*), water pennywort (*Hydrocotyle ranunculoides*), bladderworts (*Utricularia* spp.), water hyacinth (*Eichhornia crassipes*), and water lettuce (*Pistia stratiotes*). (See Plate 12.)

Tidal Swamps

Tidal swamps are located along the uppermost reaches of tidally influenced rivers and between the upland or nontidal forested wetlands (e.g., hydric hammocks) and tidal fresh marshes. Of all the coastal marshes, the tidal swamps are perhaps the least understood. Dominant plant species are much the same as in the neighboring nontidal swamps, but their composition probably differs. (See Plate 13.)

Dominant trees consist of bald cypress, water tupelo (*Nyssa aquatica*), black gum or swamp tupelo (*Nyssa sylvatica* var. *biflora*), sweet bay (*Magnolia virginiana*), water locust, red maple, elms (*Ulmus* spp.), ashes (*Fraxinus* spp.), cabbage palm, river birch (*Betula nigra*), black willow (*Salix nigra*), water hickory (*Carya aquatica*), ironwood (*Carpinus caroliniana*), loblolly pine, and sweet gum (*Liquidambar styraciflua*). Other trees that may also occur less commonly are sycamore (*Platanus occidentalis*), persimmon (*Diospyros virginiana*), Atlantic white cedar (*Chamaecyparis thyoides*), and tulip poplar (*Liriodendron tulipifera*). Locally, Atlantic white cedar may codominate with sweet bay. Associated shrubs include buttonbush, willows (*Salix* spp.), bluestem palmetto (*Sabal minor*), wax myrtle, groundsel bush, smooth alder, red bay (*Persea borbonia*), southern red cedar, and, rarely, titi (*Cyrilla racemiflora*). Common vines may be present, such as climbing hempweed, pepper-vine (*Ampelopsis arborea*), and poison ivy. The herbaceous layer may consist of sand cordgrass, royal fern, marsh fern, sawgrass, sedges, umbrella sedges, lizard's tail (*Saururus cernuus*), Asiatic dayflower, arrow arum, arrow-leaved tearthumb (*Polygonum sagittatum*), rice cutgrass, water hemlock, large-fruit beggar-ticks (*Bidens coronata*), bur marigold, southern blue flag, butterweed, and asters.

Tidal Flats

Tidal flats occur throughout the coastal zone from marine waters to tidal fresh waters. They are nearly level areas largely devoid of macrophytic vegetation located between permanently flooded shallows and tidal marshes and swamps. (See Plate 14.) In more saline areas, some flats may be covered by macroscopic algae, such as sea lettuce (*Ulva lactuca*) and other green algae (*Enteromorpha* spp.), or by microscopic diatoms. Isolated clumps of smooth cord-grass may occasionally be found at the highest elevations on the flats. In South Florida, red mangrove seedlings often can be observed growing out in the flats, extending the mangrove swamps seaward. In brackish and tidal freshwater areas, low-growing plants may be locally common on tidal flats. These plants include eastern lilaeopsis, pygmyweed (*Crassula aquatica*), mud plantain (*Heteranthera reniformis*), and false pimpernel (*Lindernia dubia*). The majority of flats, however, appear as large expanses of mud and/or sand at low tide. At the lowest levels, nearest the water, aquatic bed plants may be exposed by the lowest annual tides, especially by strong offshore winds in areas of low tidal amplitude.

Coastal Aquatic Beds

Aquatic beds are repesented by three types of plants: floating-leaved rooted, free-floating, and submerged (underwater). In addition, some typically emergent plants, such as alligatorweed, golden club, maiden-cane, and torpedo grass, may form dense floating mats. Aquatic beds often become established in shallow waters adjacent to coastal wetlands, in pools and ponds within the wetlands, and in man-made impoundments.

As with coastal wetland vegetation, the composition of the aquatic beds dramatically changes from marine waters to tidal fresh waters, with a corresponding increase in diversity as well. In marine waters and high-salinity estuarine bays, the beds may be dominated by eel-grass (*Zostera marina*), widgeon-grass, turtle-grass (*Thalassia testudinum*), manatee-grass (*Cymo-docea filiformis*), shoal-grass (*Halodule wrightii*), and Engelmann's sea-grass (*Halophila engelmannii*). With decreased salinity in upstream coastal rivers, widgeon-grass remains an important brackish aquatic bed plant, while the others are replaced by several species, including sago pondweed (*Potamogeton pectinatus*), clasping-leaved pondweed or redhead-grass (*Potamogeton perfoliatus*), horned pondweed (*Zannichellia palustris*), naiads (*Najas guadalupensis* and *N. marina*), and Eurasian water milfoil (*Myriophyllum spicatum*). Slightly brackish waters may have beds of wild celery (*Vallisneria americana*), waterweeds (*Elodea* spp.), coontail (*Ceratophyllum demersum*), leafy pondweed (*Potamogeton foliosus*), baby pondweed (*Potamogeton pusillus*), and duckweeds (*Lemna* spp. and *Spirodela polyrhiza*); yet these species are usually more prominent in tidal fresh waters. Dominant aquatic species in tidal fresh waters also include water shield (*Brasenia schreberi*), fanwort (*Cabomba caroliniana*), water hyacinth, water lettuce, floating-hearts (*Nymphoides aquaticum*), water pennywort, American frog-bit, various pondweeds (*Potamogeton* spp.), bladderworts, mosquito-fern (*Azolla caroliniana*), water lilies (*Nymphaea* spp.), spatterdock, pickerelweed, arrowheads (*Sagittaria* spp.), water-milfoils (*Myriophyllum* spp.), and marsh mermaid-weed (*Proserpinaca palustris*). Introduced (non-native) aquatic plants have established themselves in southern coastal waters, including Eurasian water-milfoil, water hyacinth, curly pondweed (*Potamogeton crispus*), South American elodea (*Egeria densa*), hydrilla (*Hydrilla verticillata*), and water chestnut (*Trapa natans*).

Plate 1. Coastal wetlands typically form behind barrier islands (*above:* Capers and Dewees Islands, South Carolina) and along tidal rivers (*below:* Combahee River, South Carolina). Note the many impounded marshes along the Combahee—former rice fields now managed for waterfowl hunting.

(A)

Plate 2. Some easily seen plant adaptations for life in salt water or saline soils are: (A) fleshy leaves and young stems (Saltwort, *Batis maritima*), (B) fleshy stems and reduced leaves (Glassworts, *Salicornia* ssp.), and (C) fleshy leaves and salt glands on leaf stalk (White Mangrove, *Laguncularia racemosa*).

(B)   (C)

Plate 3. Salt and brackish marsh species have successfully invaded this former loblolly pine (*Pinus taeda*) forest on Virginia's Eastern Shore. This is stark evidence of the impact of rising sea level on coastal vegetation.

Plate 4. Mangrove swamp along the Intercoastal Waterway in Highland Beach, Florida. Note that Australian pine (*Casuarina equisetifolia*) is invading the swamp.

Plate 5. The arching prop roots of red mangrove (*Rhizophora mangle*, on left) distinguish it from black mangrove (*Avicennia germinans*, on right) which has erect pencillike pneumatophores arising from its underground roots.

Plate 6. Dwarf red mangroves in the Everglades.

Plate 7. Low marsh dominates salt marshes along the South Atlantic coast, especially in South Carolina and Georgia.

Plate 8. Smooth cordgrass (*Spartina alterniflora*) characterizes the low marsh but is also widespread in parts of the high marsh.

Plate 9. Salt flats occur within the high marsh where soil salinities may be more than three times that of sea water. Only the most salt-tolerant species colonize these sites.

Plate 10. Black needlerush (*Juncus roemerianus*) often forms nearly monotypic stands in the more saline brackish marshes and also occurs along the upper edges of salt marshes.

Plate 11. Many of Louisiana's brackish marshes are dominated by salt meadow cordgrass (*Spartina patens*). An egret stalks the shallows for fish in this one.

Plate 12. Clumps of giant cutgrass (*Zizaniopsis miliacea*) are evident along this tidal fresh marsh.

Plate 13. Water tupelo (*Nyssa aquatica*) is among several tree species common in tidal swamps.

Plate 14. Tidal flats are exposed at low tide. They are most extensive in regions of high tidal range.

Identification of Coastal Wetland and Aquatic Plants

To identify tidal wetland and aquatic plants common in the Southeast, seven keys are provided. Because plants flower at different times during the growing season and flowers may not be present at the time of observation, vegetative characteristics are emphasized in the keys, except Key F for herbs in flower. In all of the keys, the use of technical terms has been minimized. A glossary defining technical terms is provided near the end of this book, and a scale to aid in measurements is included inside the back cover.

How to Use the Keys

A general key to Keys A–F is presented to guide you to the appropriate plant identification key. This "Key to Subsidiary Keys" separates plants first on the basis of habitat (i.e., whether growing in permanent open water or in periodically flooded areas), then on life form (e.g., woody or herbaceous plants), directing you to the pertinent life form key to begin plant identification. Each key consists of a series of couplets, for example, 1 and 1, 2 and 2, 3 and 3, and so on. Each couplet contains contrasting statements about the plant in hand. As you work down through the keys, you will eventually reach a couplet that references pages where the plant may be illustrated and described. Review appropriate illustrations, locate the one resembling the plant in hand, and then read the description. Reading the description is often necessary to identify a specific plant. In the plant descriptions, you will notice the "Similar species" subheading. It is important to read this section to distinguish the illustrated and described plant from related species or other plants resembling it or having a distinctive feature (e.g., prickles) in common with it. If the plant you have does not meet the description in the text or resemble the illustration, you will need to go through the key again, carefully checking characteristics. If the plant in question is not illustrated or covered under the "Similar species" heading in the plant descriptions, use other field guides or botanical taxonomic manuals to identify the plant (see References), or send the specimen to a botanist at a local herbarium for identification.

In the description and illustration section, plants are organized first by their typical habitat (i.e., salt and brackish coastal waters, tidal fresh coastal waters, salt and brackish coastal wetlands, and tidal fresh coastal wetlands) and then within these groups by vegetative characteristics. Recognize that many species occur in both brackish and tidal fresh wetlands; if you do not find a species under one habitat, check the other. In general, I have listed such species at the end of the group of plants with similar characteristics, under the heading "See also." This should help cross-reference the more wide-ranging species.

Overview of Plant Characteristics

Before using the keys or reading the plant descriptions, it may be useful to review some of the more important features of wetland and aquatic plants discussed and illustrated below.

Life Form

Life form relates to the growth form of a plant. Five life forms are generally recognized in tidal wetlands and coastal waters: (1) *aquatic plants*, (2) *emergent plants*, (3) *shrubs*, (4) *trees*, and (5) *vines*. Aquatic plants grow in permanently flooded waters, either free-floating, submerged (underwater), or with floating leaves at the surface and stems rooted in the underlying substrate. Emergent plants are erect, herbaceous (nonwoody) plants that have all or part of their stems and leaves standing above the water's surface or grow on the surface of intertidal soils. They can be divided into four general subtypes: (1) ferns, (2) grasses and grasslike plants, (3) fleshy flowering herbs, and (4) nonfleshy flowering herbs. Shrubs are woody plants less than 20 feet in height, usually with multiple stems, but also including saplings of tree species. Trees are woody plants 20 feet or greater in height, having a single main stem (trunk). Vines are woody or herbaceous plants that climb or twine around the stems of other plants.

Leaf Types

Leaves may be conspicuous as in most wetland plants or reduced to spines or scales as in Russian thistle and glassworts, respec-tively. *Simple leaves* consist of a single (undivided) blade, whereas *compound leaves* are divided into two or more distinct and separate parts called leaflets that can be individually removed. *Lobed leaves* are simple leaves that have shallow or deep indentations forming lobes but are not divided into separate parts. *Entire leaves* have smooth or wavy margins, lacking teeth, although they may be fringed with minute hairs (cilia). *Toothed leaves* have margins indented by fine or coarse teeth or have scalloped edges (rounded teeth). Leaves also take on a variety of shapes, including threadlike, grasslike, linear, lance-shaped, egg-shaped, spoon-shaped, heart-shaped, arrowhead-shaped, and sword-shaped. *Fleshy-leaved plants* have succulent leaves that are thickened and somewhat fleshy in texture. Leaves are attached to the stem in different ways. *Sessile leaves* are directly joined to the stem, without a stalk (petiole). *Petioled leaves* are connected to the stem by a stalk. (See Figure 4.)

Leaf Arrangements

Leaves are arranged on plants in four basic ways: (1) basal, (2) opposite, (3) whorled, and (4) alternate. *Basal leaves* grow directly from a short stem at or below the ground surface. When leaves grow in pairs across from each other along a stem occurring above the ground surface, they are *oppositely arranged*. If the leaves grow in clusters of three or more around the stem, they are *whorled*. When leaves grow singly on the stem and vary in position from one side to the next up the stem, they are *alternately arranged*.

Flower Types and Arrangements

At first glance, flowers are either regularly or irregularly shaped. *Regular flowers* are radially symmetrical with distinct petals or petallike parts surrounding the center of

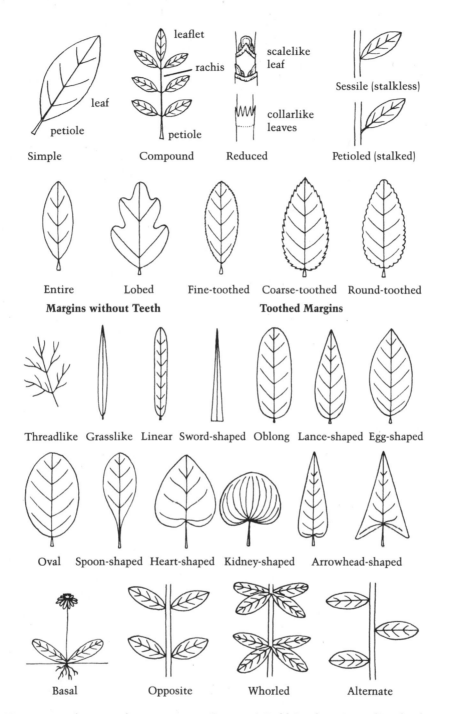

Figure 4. Leaf types and arrangements. (Source: *A Field Guide to Coastal Wetland Plants of the Northeastern United States*, Tiner 1987)

the flower. Each petal or petallike part is alike in shape, size, and color. *Irregular flowers* are not radially symmetrical, and their petals or petallike parts are not alike but differ in shape, size, and/or color. *Tubular flowers* have petals fused into lobes that are joined at the base forming a tube. Some tubular flowers have distinct upper and lower lobes called lips. Others have lobes that are deeply divided and petallike. Inconspicuous flowers do not have petals, or their petals or petallike parts are so small that they cannot be readily observed without magnification. (See Figure 5.)

Flowers are arranged in several ways: Many grow singly along the stem, whereas others occur in clusters of various types. The latter include heads, panicles, racemes, spikes, and umbels among others. A *head* is a rounded or flat-topped cluster of sessile flowers, as found in asters and goldenrods. A *panicle* is a highly branched inflorescence with a central axis, as present in many grasses. A *raceme* is an unbranched elongated inflorescence with lateral flowers on short stalks. A *spike* is a type of raceme with sessile flowers. Some plants possess a single terminal spike, and others have numerous small spikes called spikelets branching from a central axis or side branches. An *umbel* is an inflorescence with several branches arising from the end of a peduncle (flowering stalk), as found in water parsnip.

Distinguishing among Grasses, Sedges, and Rushes

Many people have difficulty separating grasses, sedges, and rushes from one another. It is, therefore, useful to review the general differences between them. (See Figure 6.)

Grasses. Grasses have jointed (swollen nodes), hollow stems (sometimes pithy) that are round in cross-section. Leaves are distinctly two-ranked and connected to the stem by open sheaths. Leaves possess a ligule (membranous or hairy appendage) at the junction of the leaf blade and leaf sheath which encircles the stem. Inconspicuous flowers are borne in spikelets, and each flower consists of two glumes (bracts) and one or more florets (each of which has two different bracts, the lemma and the palea, and stamens and pistil). Fruits are grainlike seeds covered by two papery scales.

Sedges. Sedges have solid, triangular stems for the most part; notable roundish-stem exceptions include soft-stemmed bulrush, Gulf coast spike-rush, and three-way sedge. Leaves are three-ranked with closed sheaths. Inconspicuous flowers are borne in the axils of overlapping scales, which may collectively resemble a bud. Each flower consists of a single pistil with two or three stigmas and one to three stamens. Fruits are lens-shaped or three-angled nutlets called *achenes.*

Rushes. Rushes have solid stems, mostly round in cross-section. No ligule is present. Most have few leaves. They have regular flowers with three sepals, three petals, three or six stamens, and a fruit capsule bearing numerous small seeds. Fruit capsules are usually three-valved but sometimes one-celled. Rushes can be easily separated from grasses and sedges by their multiseeded capsules, because grasses and sedges have only one seed per flowering scale.

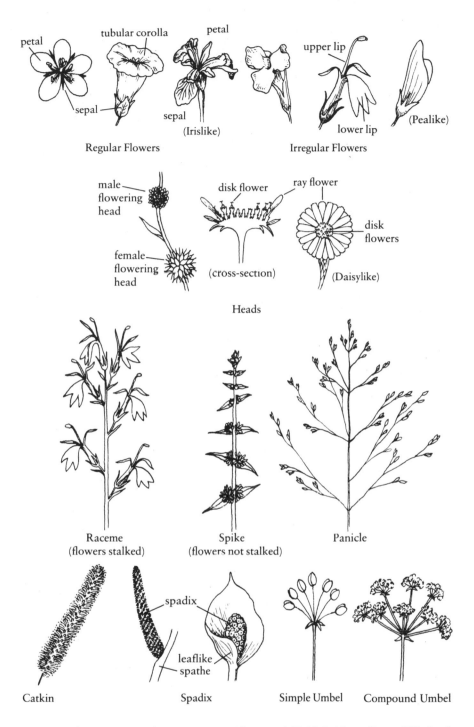

Figure 5. Flower types and arrangements. (Source: *A Field Guide to Coastal Wetland Plants of the Northeastern United States,* Tiner 1987)

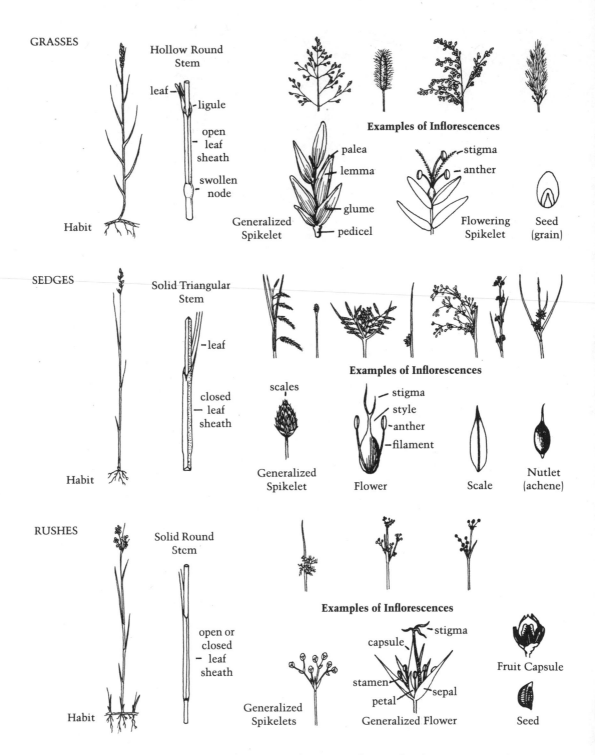

GRASSES

Habit

Hollow Round Stem

leaf — ligule

open leaf sheath

swollen node

Generalized Spikelet

palea
lemma
glume
pedicel

Examples of Inflorescences

stigma
anther

Flowering Spikelet

Seed (grain)

SEDGES

Habit

Solid Triangular Stem

leaf

closed leaf sheath

scales

Generalized Spikelet

Examples of Inflorescences

stigma
style
anther
filament

Flower

Scale

Nutlet (achene)

RUSHES

Habit

Solid Round Stem

open or closed leaf sheath

Generalized Spikelets

Examples of Inflorescences

capsule
stigma
stamen
petal
sepal

Generalized Flower

Fruit Capsule

Seed

Figure 6. Distinguishing characteristics of grasses, sedges, and rushes. (Source: *Field Guide to Nontidal Wetland Identification,* Tiner 1988)

ARTIFICIAL KEYS FOR COASTAL WETLAND AND AQUATIC PLANT IDENTIFICATION

Key to Subsidiary Keys

1. Plant growing in permanently flooded (open-water) area 2

    2. Plant typically submerged, free-floating or with floating leaves, and not standing erect out of the water .... Aquatic Plant (Key A)*

    2. Plant standing erect above the water's surface 3

        3. Plant herbaceous (nonwoody) .... Emergent Plant (Key B)*

        3. Plant woody 4

            4. Plant less than 20 feet tall, usually with multiple stems or growing in clumps .... Shrub (Key C)

            4. Plant 20 feet or taller (includes shorter saplings with a single stem or trunk) .... Tree (Key D)

1. Plant growing in a wetland that is not permanently flooded 5

    5. Plant herbaceous (nonwoody) 6

        6. Plant free-standing and self-supporting .... Emergent Plant (Key B)*

        6. Plant growing on and supported by other plants .... Vine (Key E)*

    5. Plant woody 7

        7. Plant free-standing and self-supporting 8

            8. Plant less than 20 feet tall, usually with multiple stems or growing in clumps .... Shrub (Key C)

            8. Plant 20 feet or taller (includes shorter saplings with a single stem or trunk) .... Tree (Key D)

        7. Plant growing on and supported by other plants .... Vine (Key E)

Key A. Aquatic Plants

1. Plants of salt and brackish coastal waters .... Pages 41–49

1. Plants of tidal fresh coastal waters 2

    2. Plant free-floating .... Pages 52–55

    2. Plant rooted in underlying substrate 3

        3. Leaves floating on surface .... Pages 56–61

        3. Leaves submerged .... Pages 62–65

*Note: An illustrated key to flowering herbs (Key F) is presented to facilitate identification of herbaceous plants in flower. This key includes floating and floating-leaved herbs, emergent herbs, and herbaceous vines.

Key B. Emergent Plants

1. Plant a fern	Page 68 (salt/brackish species) or pages 154–159 (tidal fresh species)
1. Plant not a fern	2
2. Plant a grass or grasslike	3
3. Plant a grass (Gramineae)	Pages 70–84 (salt/brackish species) or pages 160–169 (tidal fresh species)
3. Plant grasslike	4
4. Plant a sedge or a rush	5
5. Plant a sedge (Cyperaceae)	Pages 85–91 (salt/brackish species) or pages 170–184 (tidal fresh species)
5. Plant a rush (Juncaceae)	Pages 92–93 (salt/brackish species) or pages 185–187 (tidal fresh species)
4. Plant other grasslike	Page 94 (salt/brackish species) or pages 188–189 (tidal fresh species)
2. Plant not grasslike	6
6. Plant fleshy	Pages 95–111 (salt/brackish species) or pages 190–197 (tidal fresh species)
6. Plant not fleshy	7
7. Leaves all or mostly basal	Pages 112–114 (salt/brackish species) or pages 198–203 (tidal fresh species)
7. Leaves arranged along stem	8
8. Stem covered with prickles	Pages 204–205
8. Stem not prickly	9
9. Leaves simple	10
10. Leaf margins entire	11
11. Leaves alternately arranged	Pages 115–116 (salt/brackish species) or pages 206–217 (tidal fresh species)
11. Leaves oppositely arranged or in whorls	12
12. Oppositely arranged	Pages 117–121 (salt/brackish species) or pages 218–222 (tidal fresh species)
12. In whorls	Pages 223–224
10. Leaf margins toothed	13

13. Leaves alternately arranged	Pages 122–124 (salt/brackish species) or pages 225–231 (tidal fresh species)
13. Leaves oppositely arranged	Pages 125–129 (salt/brackish species) or pages 232–239 (tidal fresh species)
9. Leaves compound	14
14. Leaves alternately arranged	Pages 240–243
14. Leaves oppositely arranged	Pages 244–245

Key C. Shrubs

1. Leaves fleshy	Pages 130–133
1. Leaves not fleshy	2
2. Thorns present	Pages 246–247
2. Thorns absent	3
3. Leaves evergreen	4
4. Leaves broad	Pages 134–137; 144–147 (salt/brackish species) or pages 248–249 (tidal fresh species)
4. Leaves needlelike or scalelike	Page 138 (salts/brackish species) or pages 267–268 (tidal fresh species)
3. Leaves deciduous	5
5. Leaves compound	Page 139 (salt/brackish species) or pages 250–252 (tidal fresh species)
5. Leaves simple	6
6. Leaf margins entire	7
7. Leaves alternately arranged	Pages 140–141 (salt/brackish species) or pages 253–255 (tidal fresh species)
7. Leaves oppositely arranged	Pages 256–257
6. Leaf margins toothed	8
8. Leaves alternately arranged	Page 142 (salt/brackish species) or pages 258–263 (tidal fresh species)
8. Leaves oppositely arranged	Page 264

Key D. Trees

Key E. Vines

Key F. Illustrated Key to Flowering Herbs

1. Floating or floating-leaved plant 2

   2. Flowers composed of three or more separate petals 3

      3. Flowers 1 inch wide or larger

Yellowish (to 10 inches wide): Water Lotus (p. 208)

White (to 6 inches wide): White Water Lily (p. 58)

Yellow (to 4 inches wide): Banana Water Lily (p. 58)

Yellow (to 2½ inches wide): Spatterdock (p. 194)

White (to 1½ inches wide): Arrowheads (pp. 190, 198)

Purplish (to 1¼ inches wide): Water Shield (p. 58)

      3. Flowers less than 1 inch wide

Yellow (to ¾ inch wide) with four to seven petals: Seedboxes (pp. 210, 211)

White (to ¾ inch wide): Fanwort (p. 52)

Not illustrated. White (about ¾ inch wide) with five wrinkly-margined petals: Big Floating-heart (p. 58)

White (less than ¾ inch wide): American Frog-bit (p. 56)

Whitish green (very small): Water Pennywort (p. 60)

Reddish (to ¼ inch wide) with four petals: Water Purslane (p. 222)

Pink (less than ⅛ inch wide) with four petals: Water Smartweed (p. 58)

Purplish (¾–1 inch wide) with three or four petals: Water Shield (p. 58)

Greenish to brownish (½ inch long) with three sepals (no petals): Marsh Mermaid-weed (p. 56)

   2. Flowers clustered in dense heads, or flowers tubular (petals at least partly fused at base) 4

      4. Flowers in dense heads (less than ½ inch wide or long)

White: Alligatorweed (p. 56);   pipeworts (p. 200);   pondweeds (pp. 44, 46)

4. Tubular Flowers

Bluish to purplish (to 2½ inches wide): Water Hyacinth (p. 54)

Yellow (¾ inch long): Bladder-worts (p. 54)

Bluish (to ½ inch wide): Kidney-leaf Mud Plantain (p. 196)

Yellow (to ½ inch wide): Water Star-grass (p. 196)

White (¼ inch wide): Coastal Water-hyssop (p. 110)

6. Flowers larger than 1 inch wide

Rose to lavender (to 3¾ inches wide): Salt Marsh Morning Glory (p. 150)

White, pink, or purplish (to about 2 inches wide): Hedge Bindweed (p. 288)

Bluish, rose, or white (more than 1 inch wide): Leather-flower (p. 286)

6. Flowers less than 1 inch wide

Pink or white headlike clusters: Climbing Hempweed (p. 287)

Yellow (less than 1 inch long): Deer Pea (p. 148)

Purplish or brownish (about ½ inch long): Ground-nut (p. 285)

Greenish yellow (to about ¼ inch long): Wild Yam (p. 288)

Greenish white (less than ⅓ inch wide): Climbing Milkweed (p. 149)

White or yellowish (to ⅓ inch long): Dodders (p. 284)

7. Flowers larger than 3 inches wide

White (to 12 inches wide): Southern Swamp Lily (p. 200)

Yellowish (to 10 inches wide): Water Lotus (p. 208)

White or pink (to 6 inches wide): Rose Mallow (p. 122)

White (to 4 inches wide): Marsh Spider Lily (p. 206)

Bluish to purplish (to 5 inches wide): Southern Blue Flag (p. 202)

Yellow (to 4 inches wide): Yellow Flag (p. 202)

7. Flowers up to 3 inches wide 8

8. Flowers daisylike, composed of petallike rays surrounding a central disk

Yellow (to 2½ inches wide): Bur Marigold (p. 232)

Yellow (less than l inch wide): Butterweed (p. 192)

Blue to violet (to 1¼ inches wide): New York Aster (p. 208)

Not illustrated. Bluish to lavender (less than 1 inch wide): Coastal Plain Aster (p. 104)

White, blue, or pale purple (to 1 inch wide): Perennial Salt Marsh Aster (p. 104)

Not illustrated. White (less than 1 inch wide): Lowland White Aster (p. 104); Small White Aster (p. 228)

Not illustrated. White, pink, or light purple (less than 1 inch wide): White Boltonia (p. 228)

8. Flowers not daisylike 9

9. Flowers irislike

Yellow: Yellow Flag (p. 202)

Bluish to purplish: Southern Blue Flag (p. 202)

9. Flowers not irislike 10

10. Flowers larger than 1 inch wide or long 11

11. Flowers with separate petals, not noticeably fused at base

Pink, rarely white (to 2½ inches wide): Perennial Salt Marsh Pink (p. 118)

Pink, rarely white (to 2½ inches wide): Seashore Mallow (p. 124)

Pink, rarely white (to 1 ½ inches wide): Annual Salt Marsh Pink (p. 120)

Yellow (to 2½ inches wide): Spatterdock (p. 194)

Yellow (to 2½ inches wide): Seedboxes (p. 210)

Yellow (to 1¾ inches wide): Partridge Pea (p. 240)

White (to 1½ inches wide): Arrowheads (pp. 191, 198)

Not illustrated. Lavender or white with five or six petals: Catchfly Gentian (p. 118). Greenish yellow with five petals: Mangrove Mallow (p. 122)

11. Flowers tubular with petals fused at base, or flowers two-lipped (upper and lower lobes)

Bluish to purplish (to 2½ inches wide): Water Hyacinth (p. 54)

Bluish to purplish (to about 1 inch long): Elongated Lobelia (p. 226)

Red (to 1⅕ inches long): Cardinal Flower (p. 226)

Creamy white to pinkish purple (to 1⅛ inches long): American Germander (p. 126)

10. Flowers up to 1 inch wide or long 12

12. Flowers inconspicuous, apparently lacking distinct petals or lobes 13

13. Salt and brackish marsh plant

Green: Marsh Orach (p. 98); Lamb's-quarters (p. 100); Sea Blite (p. 102); Russian Thistle (p. 102)

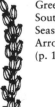 

Greenish: Southern Seaside Arrow Grass (p. 108)

Greenish to yellow-green: Water Hemp (p. 115)

Green to brown: Narrow-leaved Cattail (p. 94)

White: Bloodleaf (p. 117)

13. Slightly brackish and tidal fresh marsh plant

 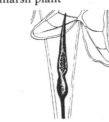 

Yellow:
Sweet Flag (p. 188); Golden Club (p. 190); Arrow Arum (p. 192)

Green: Swamp Dock (p. 214); Giant Ragweed (p. 232)

Greenish to yellow-green: Water Hemp (p. 115)

Greenish: False Nettle (p. 238)

Green or greenish white: Clearweed (p. 238)

Greenish to brownish: Marsh Mermaid-weed (p. 56)

Green to brown: Cattails (p. 188)

White: Lizard's Tail (p. 216)

12. Flowers conspicuous, often with showy petals or lobes 14

14. Flowers in headlike or ball-shaped clusters 15

15. Salt and brackish marsh plant

Yellow: Seaside Golden-rod (p. 106)

Pink: Salt Marsh Fleabane (p. 122)

Purplish or bluish: Annual Salt Marsh Aster (p. 104)

White: Thoroughworts (pp. 106, 125); Silverhead (p. 96); Common Frog-fruit (p. 126)

15. Tidal fresh marsh plant

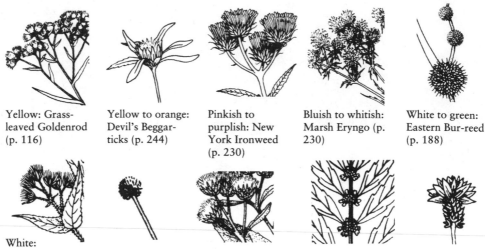

Yellow: Grass-leaved Goldenrod (p. 116)

Yellow to orange: Devil's Beggar-ticks (p. 244)

Pinkish to purplish: New York Ironweed (p. 230)

Bluish to whitish: Marsh Eryngo (p. 230)

White to green: Eastern Bur-reed (p. 188)

White:
Boneset (p. 234); Pipeworts (p. 200); Marsh Fleabane (p. 122); Water Horehound (p. 234); Alligatorweed (p. 56)

Not illustrated. Yellow: Iris-leaf Yellow-eyed Grass (p. 180)

14. Flowers not in headlike or ball-shaped clusters **16**

16. Flowers pealike

Yellow or reddish: Sensitive Joint Vetch (p. 240)

Yellow: Rattlebush (p. 139)

16. Flowers not pealike **17**

17. Flowers of milkweeds, composed of drooping lobes and erect hoods and horns

Pink: Swamp Milkweed (p. 218)   Red, orange, or reddish purple: Red Milkweed (p. 118)

17. Flowers not milkweed-type **18**

18. Flowers borne in umbels 19

19. Leaves simple

White, with ribbonlike leaves: Eastern Lilaeopsis (p. 114)

White or green, with roundish leaves: Pennyworts (p. 112)

19. Leaves compound, flowers white

Water Hemlock (p. 240)   Mock Bishop-weed (p. 242)   Water Parsnip (p. 242)

18. Flowers not borne in umbels 20

20. Flowers two-lipped (upper and lower lobes)

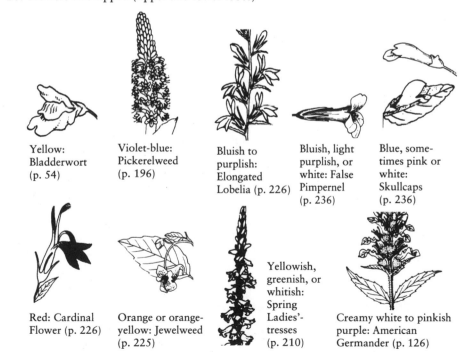

Yellow: Bladderwort (p. 54)

Violet-blue: Pickerelweed (p. 196)

Bluish to purplish: Elongated Lobelia (p. 226)

Bluish, light purplish, or white: False Pimpernel (p. 236)

Blue, sometimes pink or white: Skullcaps (p. 236)

Red: Cardinal Flower (p. 226)

Orange or orange-yellow: Jewelweed (p. 225)

Yellowish, greenish, or whitish: Spring Ladies'-tresses (p. 210)

Creamy white to pinkish purple: American Germander (p. 126)

20. Flowers not two-lipped but having distinct petals or lobes 21

21. Three petals or lobes

Pink to light purple: Asiatic Dayflower (p. 206)

Orange or orange-yellow: Jewelweed (p. 225)

White, with whorled leaves: Dye Bedstraw (p. 224)

White (about ⅗ inch wide): Awl-leaf Arrowhead (p. 198)

White (less than ⅓ inch wide): Southern Water Plantain (p. 198)

Greenish to brownish: Marsh Mermaidweed (p. 56)

21. More than three petals or lobes 22

22. Four petals or lobes

Yellow: Seedboxes (pp. 210, 211). *Not illustrated.* Common Purslane (p. 95)

Reddish: Water Purslane (p. 222)

Pink, with prickly stems: Halberd-leaved Tearthumb (p. 204)

Pink, green, or white: Virginia Knotweed (p. 214)

Pink, purplish, or white: Pink Ammania (p. 220)

White to light purple (¼ inch wide): Sea Rocket (p. 108). *Not illustrated.* Salt Marsh Loosestrife (p. 120)

White to greenish white; plant less than 4 inches tall: Pygmyweed (p. 192)

Not illustrated. White: Shade Mudflower (p. 110)

22. More than four petals or lobes 23

23. Five petals or lobes 24

24. Salt and brackish marsh plant

Lavender or purplish blue: Sea Lavenders (p. 112)

Light purple to pink: Sea Purslane (p. 95)

Pink to purplish: Seaside Gerardia (p. 110)

Pink to lavender: Sandpaper Vervain (p. 128)

Pink to white: Salt Marsh Sand Spurrey (p. 98)

White to pink: Sea Beach Knotweed (p. 108)

White to light purple: Salt Marsh Loosestrife (p. 120)

White: Seaside Heliotrope (p. 96)

Not Illustrated. Yellow: Common Purslane (p. 95)
Not Illustrated. Pinkish: Pink Purslane (p. 95)

24. Tidal fresh marsh plant

Pink to purplish: Water-willow (p. 223); Pinkweed (p. 212)

Pink, purplish, greenish white , or white: Arrow-leaved Tearthumb (p. 204) with prickly stems

Pink, purplish, or red: Marsh St. John's-wort (p. 220)

Bluish to violet: Blue Vervain (p. 128)

Yellow:
Overlooked Hedge Hyssop (p. 236); Dwarf St. John's-wort (p. 218); Seedboxes (pp. 210, 211)

Tidal fresh marsh plant (continued)

Yellow or white: Coastal
Water-hyssop (p. 110)

Greenish white or
white: Smartweeds
(p. 212)

White:
Water Pimpernel (p. 216); Mudwort (p. 202)

Not illustrated. White: Virginia Hedge Hyssop (p. 236)

23. Six or more petals or lobes

Violet-blue: Pickerel-
weed (p. 196)

Bluish: Kidney-leaf Mud
Plantain (p. 196)

Light purplish to white:
Salt Marsh Loosestrife
(p. 120)

Yellow: Water Star-grass (p. 196); Seedboxes (pp. 210, 211)

Wetland Plant Descriptions and Illustrations

The following descriptions and illustrations of coastal wetland and aquatic plants are general and intended to present more characteristics to confirm that the plant in hand is the illustrated species. The descriptions include references to scientific name, common name, plant family, life form, leaves, flowers, fruits, flowering period, habitats, wetland indicator status,* and range. (*Note*: In southernmost areas, plants may bloom earlier and longer than indicated.) In addition, other plants occurring in coastal wetlands and waters that may be confused with the described plants are listed under the heading "Similar species." More detailed plant descriptions can be found in various taxonomic manuals (see References); these and other books will prove useful for identifying plants not covered in this field guide.

Scientific names follow the *National List of Scientific Plant Names* published by the U.S. Department of Agriculture, Soil Conservation Service in 1982. Where the scientific name was recently changed, its previous name or synonym is indicated in parentheses following the current name. Scientific names are usually represented by two Latin names and one or more abbreviations and/or surnames. For example, in the name *Typha angustifolia* L., *Typha* is the genus name, *angustifolia* the specific epithet, and L. the abbreviated name of the author who first used this scientific name (in this case, Carolus Linnaeus). In general discussion, we drop the author's name and

*Wetland indicator status designated represents a plant species' frequency of occurrence in southeastern wetlands according to *National List of Plant Species that Occur in Wetlands: Southeast (Region 2)*, published by the U.S. Fish and Wildlife Service. Four major categories are recognized: (1) obligate (greater than 99 percent occurrence in wetlands), (2) facultative wetland (67–99 percent), (3) facultative (34–66 percent), and (4) facultative upland (1–33 percent). Obligate (OBL) plants almost always occur in wetlands, whereas the three facultative types of plants have varying affinities for wetlands and are also found in uplands. The obligate and facultative wetland (FACW) plants are the best vegetative indicators of wetland, whereas the facultative upland (FACU) plants are the least indicative of wetland and are generally better indicators of upland. Interestingly enough, there are many wetlands where facultative (FAC) plants (e.g., red maple, loblolly pine, and sweet gum) and occasionally facultative upland plants (e.g., American holly) predominate. By themselves, these plants may reveal very little about an area's "wetlandness," but by considering the presence, abundance, and distribution of all of the plants within an area a better assessment can be made. A positive sign (+) or a negative sign (−) following the three facultative categories indicates whether a plant is on the wetter or drier side of the category's range, respectively. For example, a FACW+ plant has a higher frequency of occurrence in wetlands (83–99 percent) than a FACW− plant (67–83 percent).

refer to the plant as the species *Typha angustifolia* or by its common name, that is, narrow-leaved cattail.

All measurements are in English units (i.e., inches and feet) because most readers are more familiar with them than with their metric equivalents.

Plant descriptions and illustrations are grouped by life form within major habitats (i.e., salt and brackish coastal waters, tidal fresh coastal waters, salt and brackish coastal wetlands, and tidal fresh coastal wetlands). Within these groups, plants are arranged by leaf characteristics for the most part, except where other traits are more diagnostic, and then alphabetically by family.

Plants of Salt and
Brackish Coastal Waters

Manatee-grass

Cymodocea filiformis (Kuetz.) Correll
[*Syringodium filiforme* Kutzing]

Manatee-grass Family
Cymodoceaceae

Description: Submerged, rooted aquatic plant of marine waters; linear, grasslike leaves (up to 20 inches long and to ⅛ inch wide) roundish in cross-section, sheathed at base, and borne in groups of two to four on short stems from rhizomes; rarely observed female and male flowers borne on short stalks from leaf sheaths; small one-seeded, somewhat egg-shaped fruit capsule.

Habitat: Various substrates in marine and estuarine waters (below low tide mark to 70-foot depth).

Wetland indicator status: OBL.

Range: Northeastern Florida south along Florida coast, west to Texas; also in Bermuda and the Caribbean.

Similar species: Shoal-grass (*Halodule wrightii*) has flattened leaf blades with two- to three-toothed tips; it is OBL.

Shoal-grass

Halodule wrightii Aschers.
[*Diplanthera wrightii* Aschers.]

Manatee-grass Family
Cymodoceaceae

Description: Submerged, rooted aquatic plant of marine waters and intertidal flats, sometimes forming extensive beds; simple, entire, thin, flat, linear, grasslike leaves (up to 8 inches long and to ⅛ inch wide; short on intertidal flats) with two- or three-pointed tips, two to four leaves per leaf sheath; inconspicuous flowers lacking petals borne from leaf sheaths, male and female flowers borne on separate plants (dioecious); roundish fruit (about ⅛ inch wide).

Habitat: Variable-salinity marine and estuarine waters to 30 feet deep (including hypersaline lagoons) and intermittently exposed tidal flats.

Wetland indicator status: OBL.

Range: North Carolina south through Florida, west to Texas; also in the Caribbean and Bermuda.

Similar species: Resembles widgeon-grass (*Ruppia maritima*), which lacks multipointed leaf tips; it is OBL.

Engelmann's Sea-grass

Halophila engelmannii Aschers.

Frog-bit Family
Hydrocharitaceae

Description: Submerged, rooted aquatic plant of marine waters; simple, entire, egg-shaped to oblong leaves (up to 12 inches long and to ¼ inch wide) with wavy margins, oppositely arranged, with two to four pairs of leaves borne at top of erect branch or stem; inconspicuous flowers lacking petals borne from leaf sheaths; oval-shaped fruit capsule bearing numerous seeds.

Flowering period: June through December.

Habitat: Sheltered marine and estuarine waters of variable salinity, often in association with Turtle-grass and Manatee-grass and in deep water to 270 feet.

Wetland indicator status: OBL.

Range: Florida Keys and along the Gulf coast to Texas; also in the Bahamas and the West Indies.

Similar species: Caribbean Halophila (*H. baillonis*) has only one pair of leaves at the top of the branches or stems; it is OBL. Johnson's Sea-grass (*H. johnsonii*) is similar to *H. baillonis*, but its leaves lack hairs; it is OBL.

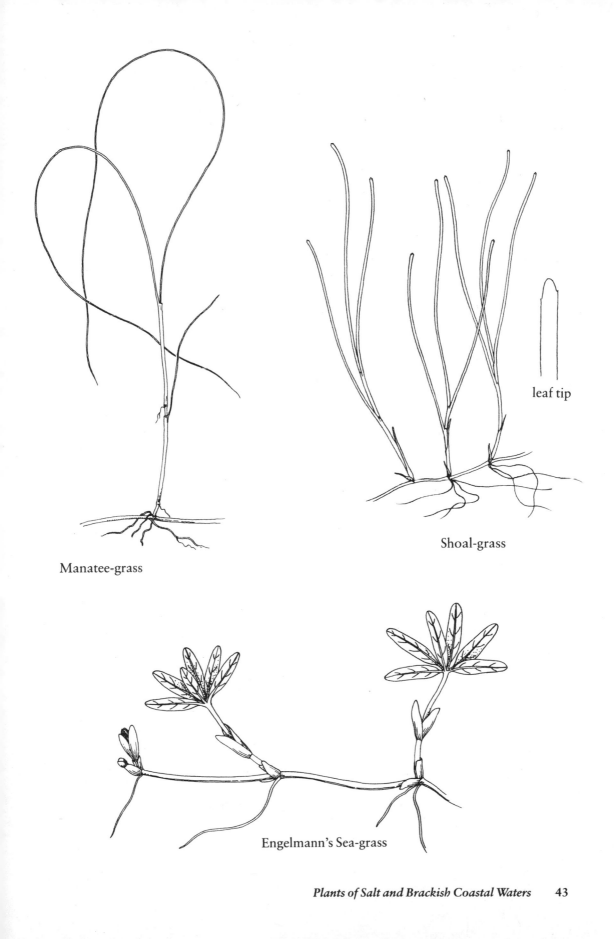

Manatee-grass

Shoal-grass

leaf tip

Engelmann's Sea-grass

Turtle-grass

Thalassia testudinum K. D. Koenig

Frog-bit Family
Hydrocharitaceae

Description: Submerged, rooted aquatic plant of marine waters, forming extensive beds; simple, entire, elongate, linear, ribbonlike, flattened leaves (up to 14 inches long and less than ½ inch wide) with rounded tips, borne in groups of two to five on short stems from rhizomes, sheathed at base, alternately arranged; minute three-"petaled" flowers borne singly from leaf axils; round, rough, warty, five- to eight-valved fruit capsule (about ¾ inch wide).

Flowering period: Mid-April into September.

Habitat: Sand, mud, or coral substrates in marine waters (from just below the low tide level to about 40 feet deep).

Wetland indicator status: OBL.

Range: Florida to Texas, south to South America; also in Bermuda and the West Indies.

Similar species: Eel-grass (*Zostera marina*) has longer leaves (to 4 feet long) that arise from branched stems; it ranges much farther north where it dominates coastal marine waters; it is OBL.

Sago Pondweed

Potamogeton pectinatus L.

Pondweed Family
Potamogetonaceae
(*Najadaceae* or *Zosteraceae*)

Description: Rooted, submerged aquatic plant; stems much branched; simple, entire, linear, septate (with cross-veins) leaves (1–4 inches long) threadlike, tapering to a long point, often with one vein; stipules fused or united to leaf forming a sheath; several whorls of minute flowers borne on spikes (½–1½ inches long) on stalks (peduncles, to 4 inches long); fruit nutlet (achene).

Flowering period: Summer.

Fruiting period: June to September.

Habitat: Brackish and tidal fresh waters; shallow fresh (calcareous) waters of lakes and slow-flowing streams.

Wetland indicator status: OBL.

Range: Quebec and Newfoundland to Alaska and British Columbia, south to Florida, Texas, and southern California.

Similar species: Other pondweeds lack septate leaves. Leafy Pondweed (*P. foliosus*) and Baby Pondweed (*P. pusillus*) have stipules that are free from leaf bases and leaves with three to five veins. *P. pusillus* usually has a pair of glands at the base of the leaves, while *P. foliosus* usually does not. Both are OBL.

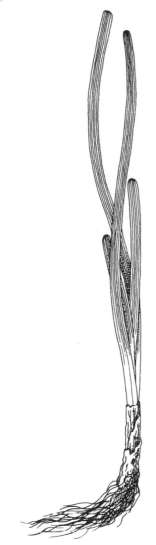

Turtle-grass

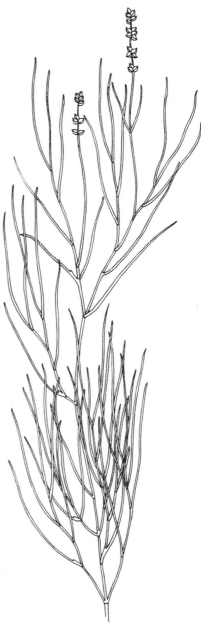

Sago Pondweed

Clasping-leaved Pondweed or Redhead-grass

Potamogeton perfoliatus L.

Pondweed Family
Potamogetonaceae
(*Najadaceae* or *Zosteraceae*)

Description: Rooted, submerged aquatic plant; stems slender, usually short, and much branched with internodes ½–1¼ inches long; egg-shaped or rounded leaves (½–3¼ inches long, usually ½–1½ inches long) with round or broad tips, bases often heart-shaped and clasping stem, with three, sometimes five, prominent nerves and several weaker ones; flowers borne in dense spikes on end of stalk (peduncle, 1¼–4¾ inches long); fruit nutlet (achene).

Flowering period: June to October.

Habitat: Brackish and tidal fresh waters; ponds and slow-moving streams.

Wetland indicator status: OBL.

Range: Newfoundland and Quebec to Ohio, south to Florida and Louisiana.

Similar species: Curly Pondweed (*P. crispus*) has wavy-margined leaves; it occurs from North Carolina north. Other *Potamogeton* in tidal fresh waters with only submerged leaves have linear leaves (*P. foliosus*, *P. pectinatus*, and *P. pusillus*). All pondweeds are OBL.

Widgeon-grass

Ruppia maritima L.

Pondweed Family
Potamogetonaceae
(*Najadaceae* or *Zosteraceae*)

Description: Rooted, submerged aquatic plant; stems simple or branched and up to 3 feet long; simple, entire, linear leaves (up to 4 inches long) threadlike, with leaf sheaths present, alternately arranged; flowers and fruits borne on stalks (⅕–12 inches long); fruit fleshy (drupe).

Flowering period: Summer.

Fruiting period: July to October.

Habitat: Saline and brackish waters, rarely tidal fresh waters, salt ponds and pools within salt marshes; inland saline waters, rarely fresh waters.

Wetland indicator status: OBL.

Range: Newfoundland to Florida and Mexico; along Pacific coast from Washington to California; inland from western New York to British Columbia.

Similar species: The leaves of Horned Pondweed (*Zannichellia palustris*) are also threadlike but oppositely arranged; its flowers and fruits are very short-stalked; it is OBL. Shoal-grass (*Halodule wrightii*) has two- or three-pointed leaf tips and two to four leaves per leaf sheath; it is OBL.

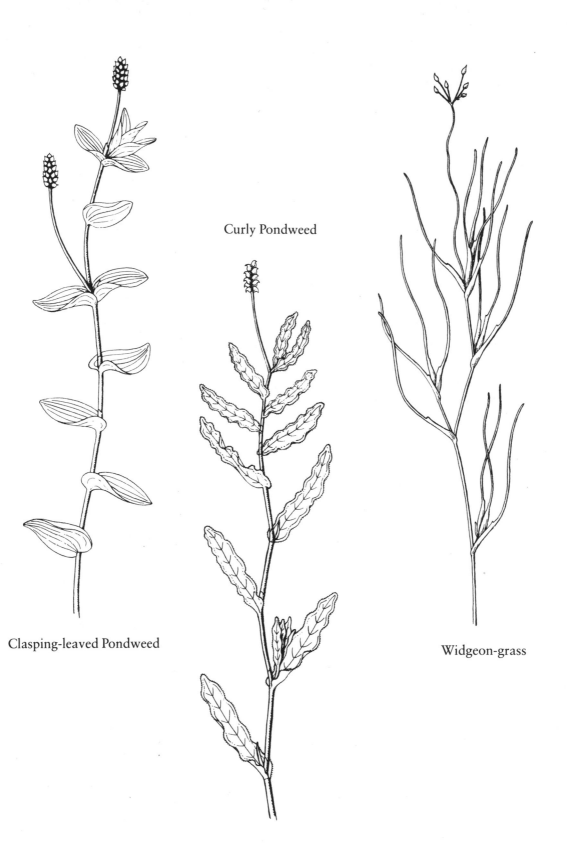

Curly Pondweed

Clasping-leaved Pondweed

Widgeon-grass

Horned Pondweed

Zannichellia palustris L.

Pondweed Family
Potamogetonaceae
(*Najadaceae* or *Zosteraceae*)

Description: Rooted, submerged aquatic plant; stems very slender, fragile, and branched (up to 20 inches long); simple, entire linear leaves (up to 4 inches long) threadlike, oppositely arranged; minute flowers borne in leaf axils enclosed by a sheath; fruit oblong nutlet (achene).

Flowering period: February to October.

Habitat: Brackish and tidal fresh waters; inland fresh and alkaline waters.

Wetland indicator status: OBL.

Range: Newfoundland and Quebec to Alaska, south to Florida, Texas, and Mexico.

Similar species: The leaves of Southern Naiad or Bushy Pondweed (*Najas guadalupensis*) are also linear and oppositely arranged but are shorter, more crowded, very finely toothed (microscopically), and not threadlike. The leaves of Widgeon-grass (*Ruppia maritima*) are also threadlike but are alternately arranged. Both are OBL.

Eel-grass

Zostera marina L.

Pondweed Family
Potamogetonaceae
(*Najadaceae* or *Zosteraceae*)

Description: Rooted, submerged aquatic plant, sometimes exposed at extreme low tides; stems slender and branched; simple, entire, linear, ribbonlike leaves (up to 4 feet long and ½ inch wide) with three to five distinct nerves; inconspicuous (hidden) flowers borne on one side of leaf enclosed within a sheath; fruit cylinder-shaped seed.

Flowering period: Summer.

Habitat: Shallow estuarine saline waters in sheltered bays and coves, occasionally tidal flats.

Wetland indicator status: OBL.

Range: Greenland and Labrador to Florida; also along the Pacific coast.

Similar species: Leaves of Wild Celery (*Vallisneria americana*) are somewhat similar, but this plant grows in slightly brackish and tidal fresh coastal waters; it is OBL.

Horned Pondweed

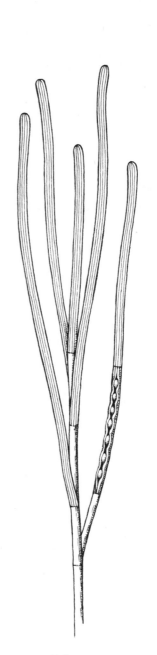

Eel-grass

*Plants of Tidal
Fresh Coastal Waters*

FREE-FLOATING AQUATICS

Coontail

Ceratophyllum demersum L.

Hornwort Family
Ceratophyllaceae

Description: Free-floating, submerged aquatic plant; stems much branched and forming large masses; compound, toothed linear leaves (²/₅– 1¹/₅ inches long), two or three times divided, arranged in five to twelve whorls; minute flowers borne singly in leaf axils; fruit nutlet (achene).

Flowering period: May through September.

Habitat: Tidal fresh waters; inland lakes and slow-flowing streams.

Wetland indicator status: OBL.

Range: Quebec to northern British Columbia, south to Florida, Texas, and California.

Similar species: Fanwort (*Cabomba caroliniana*) may occur in freshwater impoundments adjacent to tidal waters; it is a rooted aquatic plant with two types of leaves: (1) oppositely arranged compound submerged leaves divided into many linear leaflets, and (2) alternately arranged simple floating leaves (up to about 1¼ inches long); it has small white, pink, or purplish flowers (up to about ¾ inch long and wide) borne on long stalks from axils of floating leaves; it is OBL.

Big Duckweed

Spirodela polyrhiza (L.) Schleid.

Duckweed Family
Lemnaceae

Description: Free-floating, surface-water aquatic plant, often forming massive carpetlike beds on the water's surface; stem lacking; leaflike structure (thallus) broadly oval-shaped (¹/₁₀–²/₅ inch long), dark green above, purple below, with six to eighteen, usually seven, nerves and six to eighteen rootlets from the underside of the thallus; flowers in pouches (rarely seen).

Flowering period: Summer.

Habitat: Tidal fresh waters; freshwater lakes, ponds, and slow-flowing streams.

Wetland indicator status: OBL.

Range: Nova Scotia to British Columbia, south to Florida, Texas, and Mexico.

Similar species: Little Duckweed (*Lemna minor*) has only one rootlet per plant (thallus); it is OBL. Duckweeds are among the smallest of our aquatic plants.

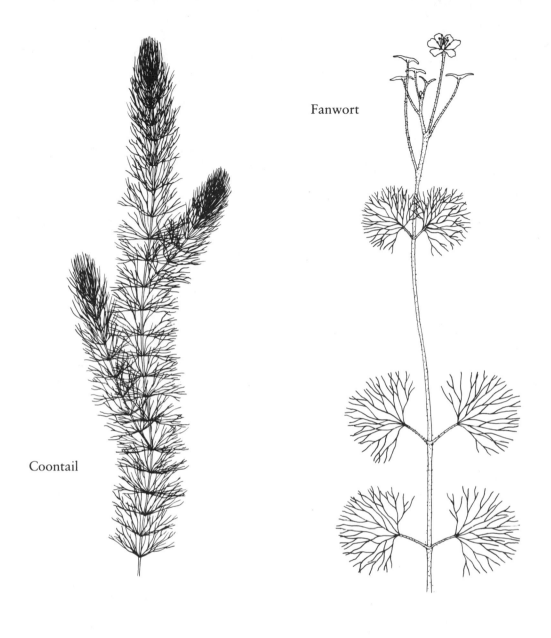

Coontail

Fanwort

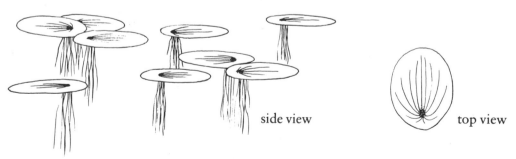

side view

top view

Big Duckweed

Common Bladderwort

Utricularia macrorhiza LeConte
[*Utricularia vulgaris* L.]

Bladderwort Family
Lentibulariaceae

Description: Free-floating, submergent aquatic plant, floating just below the water's surface; massive stems, up to 7 feet long, with many threadlike, leafy branches; rootlike leaves (up to 3 inches long) with many air-filled bladders; two-lipped yellow flowers (¾ inch long) borne on flowering stems rising above water's surface.

Flowering period: May into September.

Habitat: Tidal fresh waters; still waters of ponds and lakes, and ditches.

Wetland indicator status: OBL.

Range: New Brunswick to Alaska, south to Florida and California.

Similar species: Common Bladderwort is the common floating species. Horned Bladderwort (*U. cornuta*) and Rush Bladderwort (*U. juncea*) both have distinctive downward-pointing flower spurs (more than ⅓ inch long in the former and shorter in the latter species); both are OBL. These two species are also frequently rooted in exposed sands or peats. Zigzag Bladderwort (*U. subulata*) is an emergent species of southern brackish and tidal fresh marshes; it is a low-growing (less than 7 inches tall) delicate plant with few much reduced leaves along stem and usually two to four, two-lipped yellow flowers (less than ½ inch long); it is OBL.

Water Hyacinth

Eichhornia crassipes (Mart.) Solms.

Pickerelweed Family
Pontederiaceae

Description: Floating, somewhat fleshy aquatic plant, often forming extensive beds; simple, entire, thick, leathery or somewhat fleshy, wide egg-shaped to roundish leaves (up to 6 inches long and wide) borne on inflated stalks (to 10 inches long); numerous showy, bluish or light purplish, six-"petaled" tubular flowers (about 1–2½ inches wide) with upper middle "petal" yellow-centered, borne on a separate flowering stalk (spike, up to 15 inches long).

Flowering period: April through September.

Habitat: Slightly brackish and tidal fresh waters and muddy shores; ponds, lakes, rivers, canals, ditches, and impoundments. (*Note:* This plant is a major nuisance, clogging up many acres of waterways in the Southeast; however, it has high potential as a water-pollutant filter, if harvested.)

Wetland indicator status: OBL.

Range: Southeastern Virginia and North Carolina to Florida, west to Texas; also in California, Mexico, the West Indies, and tropical America; native of Brazil.

Similar species: Water Lettuce (*Pistia stratiotes*), another floating aquatic plant of similar habitats, is quite different in appearance; it has a mass (rosette) of grayish or dull green, hairy, ribbed, lettucelike leaves (up to 10 inches long); it is OBL.

Mosquito-fern

Azolla caroliniana Willd.

Salvinia Family
Salviniaceae

Description: Free-floating, mosslike aquatic fern (less than ½ inch wide), forming dense mats; branched stems covered by smooth, dark red to green, two-lobed leaves, overlapping and arranged in two rows; minute spore-bearing organs (sporangia) borne on lower leaf lobes.

Flowering period: June through September.

Habitat: Tidal fresh waters; nontidal waters, coastal impoundments, ponds, slow-moving streams, and exposed muds.

Wetland indicator status: OBL.

Range: Massachusetts and Wisconsin south to Florida and Texas; also in the West Indies and Mexico.

SEE ALSO American Frog-bit (*Limnobium spongia*).

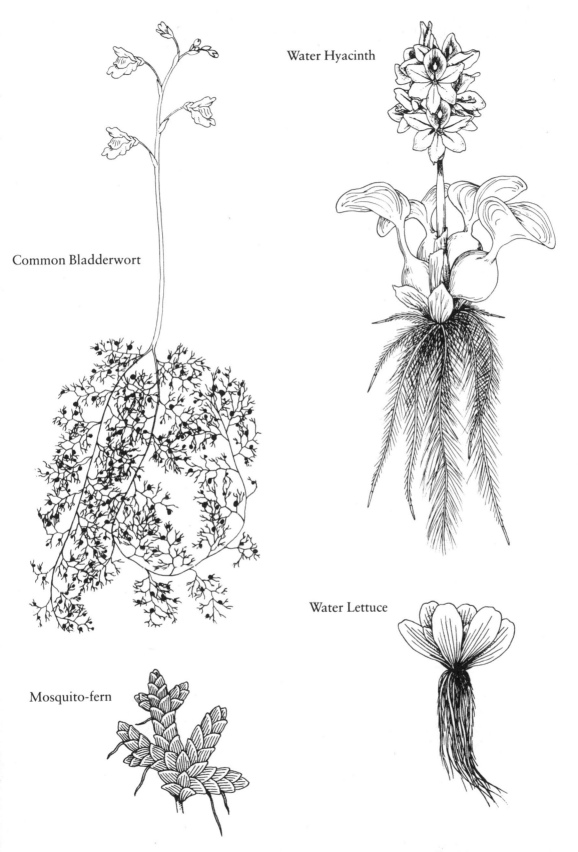

Water Hyacinth

Common Bladderwort

Water Lettuce

Mosquito-fern

ROOTED AQUATICS WITH FLOATING LEAVES

Alligatorweed

Alternanthera philoxeroides Griseb.

Amaranth Family
Amaranthaceae

Description: Rooted, floating-leaved, sometimes emergent aquatic plant, often forming extensive mats; stems mostly smooth with vertical lines; simple, entire (actually minutely toothed) linear to somewhat lance-shaped leaves (up to 5 inches long and to ¾ inch wide) stalkless or short-stalked, soft whitish hairs in leaf axils, oppositely arranged; small fragrant white flowers borne in dense headlike clusters (about ½ inch long) on long stalks (up to 3¼ inches long) from upper leaf axils and top of stem; one-seeded, smooth, bladderlike fruit.

Flowering period: April through October.

Habitat: Slightly brackish and tidal fresh water marshes and waters; nontidal marshes, rivers, lakes, ponds, streams, and ditches.

Wetland indicator status: OBL.

Range: Virginia south along the Coastal Plain to Florida, west to Texas; also in Central America; native of South America.

Marsh Mermaid-weed

Proserpinaca palustris L.

Water-milfoil Family
Haloragaceae

Description: Floating-leaved and submerged aquatic bed plant, sometimes an emergent herb; two types of leaves—(1) submerged leaves (up to 1¼ inches long), deeply dissected, often with minute spines in axils, and (2) floating or emergent leaves (up to 3¼ inches long), linear to oblong and toothed, all leaves alternately arranged; inconspicuous greenish to brownish flowers borne singly or in small clusters from upper leaf axils; fruit nutlets.

Flowering period: May through October.

Habitat: Tidal fresh waters; ponds, lakes, nontidal marshes, and wet muddy shores.

Wetland indicator status: OBL.

Range: Nova Scotia and Quebec to western Ontario and Minnesota, south to Cuba and Mexico.

Similar species: Comb-leaf Mermaid-weed (*P. pectinata*) has all of its leaves deeply divided and comblike; it is OBL. Water-milfoils (*Myriophyllum* spp.) have much-reduced emergent leaves and at least some leaves arranged in whorls; they are OBL.

American Frog-bit

Limnobium spongia (Bosc.)
    L. C. Rich ex Steud.

Frog-bit Family
Hydrocharitaceae

Description: Free-floating or rooted floating-leaved aquatic plant; simple, entire, somewhat round or kidney-shaped basal leaves (up to 3½ inches wide) often with heart-shaped bases, long-stalked (to 8 inches long), young leaves purplish below; several small three-petaled and three-sepaled white flowers (unisexual = monecious) borne singly on somewhat long stalks (to 2½ inches long) from leaf axils; somewhat roundish fleshy fruit bearing many seeds.

Flowering period: June through September.

Habitat: Tidal fresh marshes and rivers; slow-flowing rivers, nontidal marshes, ponds, lakes, ditches, and swamps.

Wetland indicator status: OBL.

Range: New Jersey south to northern Florida, west to eastern Texas, and north to southern Illinois; also in Lake Ontario and tropical America.

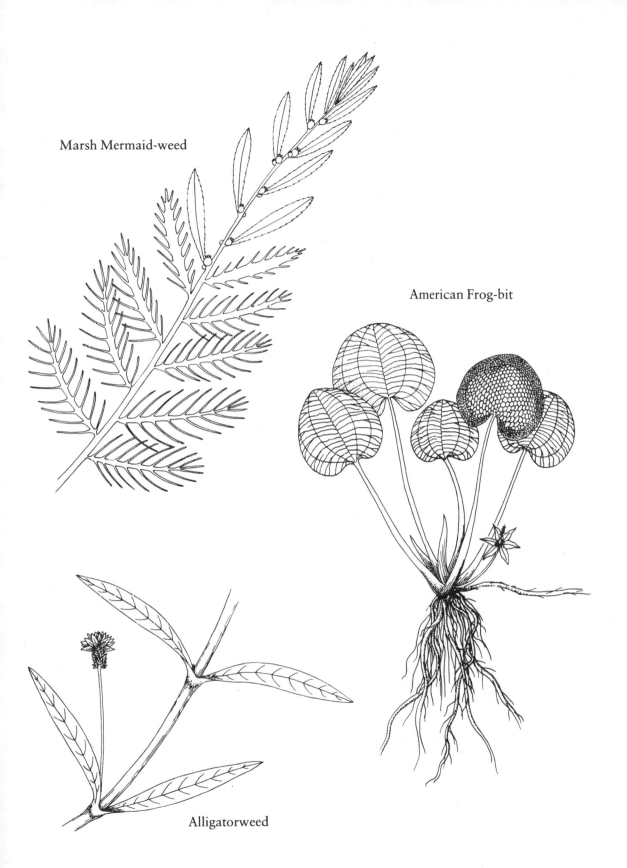

Marsh Mermaid-weed

American Frog-bit

Alligatorweed

Water Shield

Brasenia schreberi Gmel.

Water Lily Family
Nymphaeaceae

Description: Floating-leaved, rooted aquatic plant with slimy stem, up to 4 feet long; simple, entire, oval to elliptic leaves (1½– 5 inches long, ¾–3 inches wide) often slimy beneath, alternately arranged, leaf stalk attached to middle of lower leaf; small three- to four-petaled purplish flowers (¾–1¼ inches wide) borne singly on elongate stalks; small club-shaped fruits.

Flowering period: June through October.

Habitat: Tidal fresh waters; ponds, shallow lake margins, and slow-flowing rivers.

Wetland indicator status: OBL.

Range: Nova Scotia and eastern Quebec west to Minnesota and south to Florida and Texas.

White Water Lily

Nymphaea odorata Soland. in Ait.

Water Lily Family
Nymphaeaceae

Description: Rooted, floating-leaved perennial aquatic plant; elongate, branched rhizome; roundish floating leaves (up to 10 inches wide) notched at base, green above and normally purplish below, attached to rhizome by long purple to red stalk (petiole); large, showy, fragrant white (rarely pink) flower (2–6 inches wide) with many petals (seventeen to thirty-two) borne singly on a long stalk.

Flowering period: April into October.

Habitat: Tidal fresh waters; inland shallow waters of lakes and ponds.

Wetland indicator status: OBL.

Range: Newfoundland to Manitoba and northern Minnesota, south to Florida and Louisiana.

Similar species: Banana Water Lily (*N. mexicana*) may occur in slightly brackish and freshwater impoundments adjacent to tidal waters; its flowers are yellow; it is OBL. Spatterdock (*Nuphar luteum*) has erect, fleshy, heart-shaped leaves with a distinct midrib below and its flower is yellow with five to six petals; it is OBL. Narrow-leaved Yellow Pond Lily (*N. luteum* ssp. *sagittifolium*) is like Spatterdock but has floating leaves that are more than three times as long as wide; it is OBL. Big Floating-heart (*Nymphoides aquatica*) occurs in oligohaline (slightly brackish) and tidal fresh waters; it has distinctly heart-shaped to kidney-shaped leaves (to 6 inches long) that are green above and rough and purplish below and has five-"petaled" white flowers (about 1 inch wide) borne in clusters on red-spotted, elongate stalks (up to 3⅕ inches long); it is OBL.

Water Smartweed

Polygonum amphibium L.
[*Polygonum coccineum* Muhl. ex Willd.]

Buckwheat or Smartweed Family
Polygonaceae

Description: Floating-leaved, rooted aquatic plant or erect perennial herb, up to 3 feet tall; jointed stems swollen at nodes, floating and submerged stems often inflated; simple, entire leaves (up to 6 inches long) with rounded or heart-shaped to tapered bases, sheaths (ocrea) at base of leaf bearing stiff hairs, alternately arranged; many small, bright pinkish flowers (less than ⅛ inch wide) borne in dense terminal spikes (up to 1½ inches long and ¾ inch wide); nutlets.

Flowering period: June through August.

Habitat: Deep or shallow tidal waters; inland waters, nontidal marshes, and wet soils.

Wetland indicator status: OBL.

Range: Labrador and Nova Scotia to Alaska, south to Virginia, Texas, and California.

Similar species: Dense-flower Smartweed (*P. densiflorum*) occurs in southern tidal fresh waters and forested wetlands along the Coastal Plain; its sheaths (ocrea) do not have stiff hairs, its stems are reddish brown to purplish in color, its flowers are greenish white, pinkish, or green and dotted with minute glands, and its inflorescences are both terminal and axillary and are few- to many-branched; it is OBL.

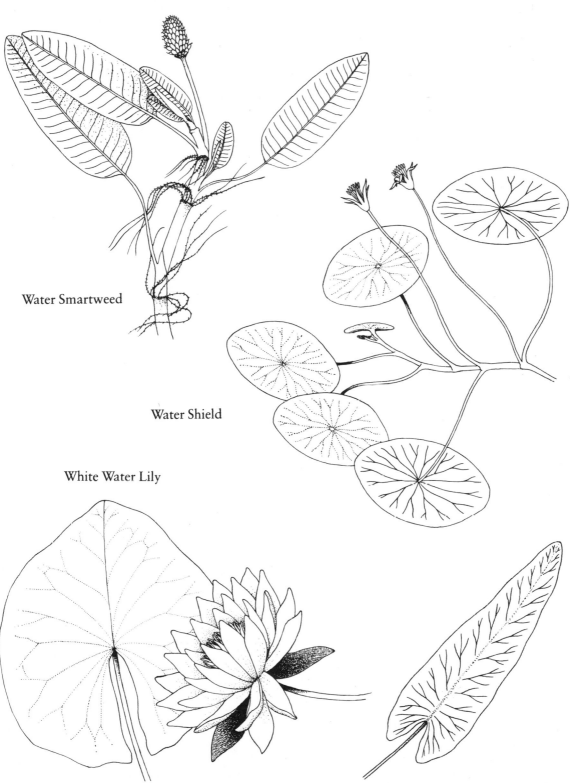

Water Smartweed

Water Shield

White Water Lily

Narrow-leaved Yellow Pond Lily leaf

Rooted Aquatics with Floating Leaves 59

Ribbonleaf Pondweed

Potamogeton epihydrus Raf.

Pondweed Family
Potamogetonaceae
(*Najadaceae* or *Zosteraceae*)

Description: Rooted, submerged, and floating-leaved aquatic plant; stems flattened, often branched; two types of leaves—(1) simple, entire, linear submerged leaves (up to 8 inches long) with five to seven nerves and lacking leaf stalk (petiole), and (2) simple, entire, egg-shaped to spoon-shaped (spatulate) floating leaves ($1\frac{1}{4}$–$2\frac{4}{5}$ inches long and $\frac{1}{3}$–$\frac{4}{5}$ inches wide) with eleven to twenty-seven nerves and with leaf stalk (petiole); flowers borne in numerous dense spikes ($\frac{2}{5}$–$1\frac{1}{5}$ inches long) on end of stalk (peduncle, 1–2 inches long); fruit nutlet (achene).

Flowering period: June through September.

Habitat: Tidal fresh waters; ponds and slow-moving streams.

Wetland indicator status: OBL.

Range: Newfoundland and Quebec to southern Alaska, south to Georgia, Iowa, Colorado, and California.

Similar species: Curly Pondweed (*P. crispus*) has curly, wavy-margined leaves. Longleaf Pondweed (*P. nodosus*) has linear to narrowly lance-shaped submerged leaves with seven to fifteen nerves. All pondweeds are OBL.

Water or Floating Pennywort

Hydrocotyle ranunculoides L.f.

Parsley Family
Umbelliferae

Description: Somewhat fleshy, floating-leaved aquatic plant, often forming dense mats; simple, thick, somewhat fleshy, round-toothed, three- to five-lobed, kidney-shaped basal leaves (up to 2 inches wide) borne on long stalks (up to 14 inches long); four to ten minute whitish-green flowers borne on a separate simple, branched terminal inflorescence (umbel, less than $\frac{1}{4}$ inch wide) with short stalks (much shorter than leaf stalks).

Flowering period: April through July.

Habitat: Margins of tidal rivers; shores of inland rivers, ditches, nontidal swamps, shallow water, and seepage areas.

Wetland indicator status: OBL.

Range: Pennsylvania south to Florida, west to Texas; also in Arizona, on the West Coast, and in tropical America.

Similar species: Other pennyworts (*Hydrocotyle* spp.) in tidal marshes have peltate leaves (with stalks attached to the center of the leaf).

SEE ALSO Water Lotus (*Nelumbo lutea*), Spatterdock (*Nuphar luteum*), Parker's Pipewort (*Eriocaulon parkeri*), and Kidney-leaf Mud Plantain (*Heteranthera reniformis*).

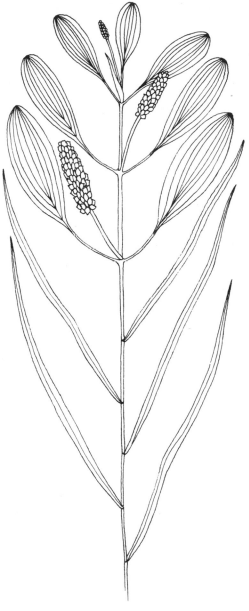

Ribbonleaf Pondweed

Water Pennywort

ROOTED AQUATICS WITH SUBMERGED LEAVES ONLY

Eurasian Water-milfoil

Myriophyllum spicatum L.

Water-milfoil Family
Haloragaceae

Description: Submerged aquatic plant forming extensive colonies; stems much branched; compound, grayish, featherlike leaves divided into twelve or more pairs of threadlike leaflets, borne in three to five whorls; minute flowers with reddish petals borne in whorled clusters of four (usually) forming a naked inflorescence; fruit capsule somewhat rounded and four-lobed.

Flowering period: April through September.

Habitat: Brackish and tidal fresh waters; lakes, ponds, sluggish streams, and impoundments.

Wetland indicator status: OBL.

Range: Southeastern Labrádor to Alaska, south to Florida, Texas, and southern California; a widespreading native of Eurasia.

Similar species: Variable Water-milfoil (*M. heterophyllum*) has variously shaped leaves ranging from toothed lance-shaped to featherlike on the same plant; it is OBL.

Waterweed

Elodea canadensis Michx.
[*Anacharis canadensis* (Michx.) Rich.]

Frog-bit Family
Hydrocháritaceae

Description: Rooted, submerged aquatic plant, sometimes floating at surface in shallow water; stems many-branched, often forming dense masses; elongate, sessile, dark green leaves usually twice as long as wide (about ³/₅ inch long and ¹/₅ inch wide), mostly drooping downward, with finely toothed margins, mostly arranged in whorls of threes; male and female flowers borne on stalks (pedicels) arising from tubular struc-

ture (spathe) in leaf axil, male stalks longer than female stalks; fruit cylinder-shaped capsule.

Flowering period: July to September.

Habitat: Tidal fresh waters; inland waters, often calcareous.

Wetland indicator status: OBL.

Range: Quebec to Saskatchewan and Washington, south to North Carolina, Alabama, Oklahoma, and California.

Similar species: Western Waterweed (*E. nuttalli*) has narrower leaves (less than ¹/₅ inch wide), and its male flowers are not borne on a long stalk; it is OBL. South American Elodea (*Egeria densa*, formerly *Elodea densa*) has longer leaves (⅘–1⅖ inches) arranged in whorls of fours to sixes; it is OBL.

Hydrilla

Hydrilla verticillata Royle

Frog-bit Family
Hydrocharitaceae

Description: Submerged, perennial aquatic bed plant; underground tubers; erect and horizontal stems; simple toothed leaves (about ½ inch long) arranged mostly in whorls of three to ten; small white flowers floating on or near the water's surface.

Flowering period: Midsummer into fall.

Habitat: Slightly brackish and fresh tidal waters; still or flowing nontidal waters and impoundments.

Wetland indicator status: OBL.

Range: Limited distribution; common in the Potomac River and other coastal rivers; native of Africa.

Similar species: Waterweeds (*Elodea* spp.) have entire leaves in whorls.

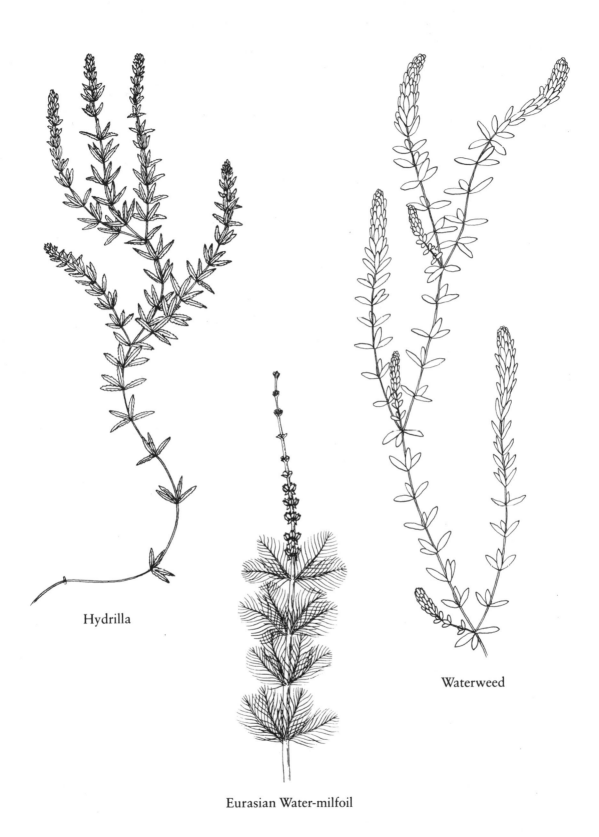

Hydrilla

Eurasian Water-milfoil

Waterweed

Wild Celery or Tape-grass

Vallisneria americana Michx.

Frog-bit Family
Hydrocharitaceae

Description: Rooted, submerged aquatic plant; stems buried in mud; simple, entire, very thin linear basal leaves (up to 7 feet long) ribbonlike; two types of flowers—male flowers borne in structure (spathe) at base of leaves and released to water's surface, female flowers borne on long stalk (peduncle) reaching water's surface; fruit cylinder-shaped capsule, peduncle coils after fertilization pulling fruit beneath water.

Flowering period: July to October.

Habitat: Tidal fresh waters, occasionally slightly brackish waters; inland waters.

Wetland indicator status: OBL.

Range: New Brunswick, Nova Scotia, and Quebec to North Dakota, south to Florida and Texas.

Similar species: Leaves of Eel-grass (*Zostera marina*) are somewhat similar, but this plant grows only in saline coastal waters; it is OBL.

Southern Naiad or Bushy Pondweed

Najas guadalupensis (Spreng.) Morong

Naiad Family
Najadaceae

Description: Rooted, submerged aquatic plant; stems very long and leafy; simple, somewhat entire (actually with twenty or more microscopic teeth along each margin), linear leaves (less than 1 inch long), dark green or olive-colored, expanded at base and sloping gradually above to a rounded or fine-pointed end, oppositely arranged; small flowers borne singly in leaf axils; purplish brown fruit enclosing straw-colored, ribbed seed.

Flowering period: August to October.

Habitat: Tidal fresh waters; inland lakes and ponds.

Wetland indicator status: OBL.

Range: Southeastern Massachusetts and New York to South Dakota and Idaho, south to Florida and Texas; also in the Pacific states.

Similar species: Spiny Naiad (*N. marina*) has conspicuously toothed leaves and may occur in brackish waters but is more common in fresh water; it is OBL.

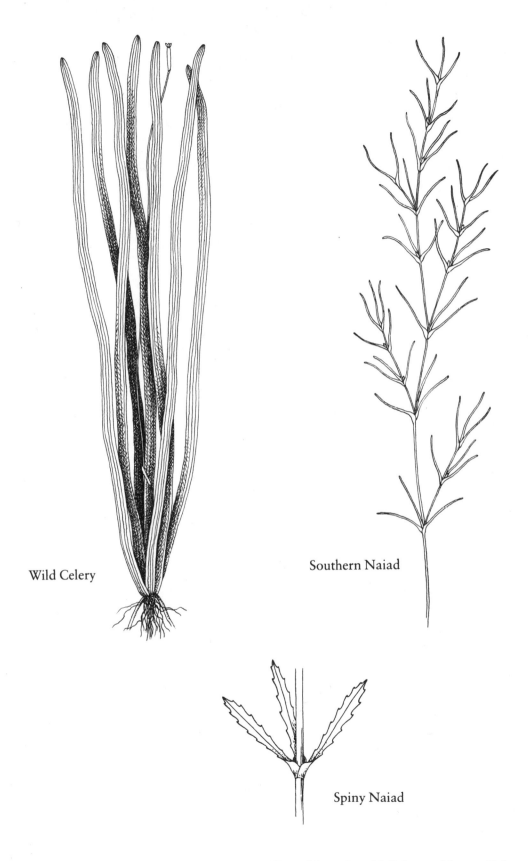

Wild Celery

Southern Naiad

Spiny Naiad

*Plants of Salt and
Brackish Coastal Wetlands*

FERNS

Coastal Leather Fern
or **Mangrove Fern**
Acrostichum aureum L.

Polypody Fern Family
Polypodiaceae

Description: Tall, leathery tropical fern, up to 7 feet high, rarely taller; thick, many-branched rhizomes; compound, thick, leathery fern leaves (up to 7 feet long and to 1 foot wide) divided into ten to fifteen pairs of leaflets (pinnae, 4–8 inches long and to 1½ inches wide) with rounded tips, veins of sterile leaflets smooth, slightly raised, and diverging at roughly 50-degree angle from midrib, leaf stalks often armed with horny spurs; separate fertile fronds with upper three to five pairs of leaflets fertile (covered by sporangia).

Habitat: Mangrove swamps and salt and brackish marshes.

Wetland indicator status: OBL.

Range: Peninsular Florida; also in West Indies and throughout tropical America.

Similar species: Inland Leather Fern (*A. danaeifolium*) occurs in brackish and freshwater marshes; it is taller (to 15 feet or more) and usually has unbranched rhizomes, its fronds usually are composed of twenty to thirty pairs of leaflets with somewhat pointed tips, its veins are often stiff-hairy and diverge at a 75-degree angle from midrib, and nearly all leaflets of the fertile fronds are covered by sporangia; it is OBL.

SEE ALSO Marsh Fern (*Thelypteris thelypteroides*).

Coastal Leather Fern

GRASSES

Salt Grass or Spike Grass

Distichlis spicata (L.) Greene

Grass Family
Gramineae (*Poaceae*)

Description: Low-growing, erect perennial grass, 8–16 inches tall; creeping rhizomes, often forming dense mats; stems stiff, hollow, and round; numerous linear leaves (2–4 inches long and most less than ⅕ inch wide) with smooth margins usually rolled inwardly and sheaths overlapping, distinctly two-ranked; terminal inflorescence (panicle, to 2½ inches long) bears one of two types of crowded (many-flowered) spikelets on separate plants (dioecious), male spikelets with eight to twelve flowers, and female spikelets usually five-flowered (three to nine).

Flowering period: June into October.

Habitat: Irregularly flooded salt marshes (often intermixed with Salt Meadow Cordgrass [*Spartina patens*] or in pure stands in wet depressions), brackish marshes, and tidal fresh marshes (less commonly).

Wetland indicator status: FACW+.

Range: New Brunswick and Prince Edward Island south to Florida and Texas; locally inland to Missouri; also along the Pacific coast.

Similar species: See Coastal Dropseed (*Sporobolus virginicus*) which has much reduced upper leaves.

Marsh Finger Grass

Eustachys glauca Chapm.
[*Chloris glauca* (Chapm.) Wood.]

Grass Family
Gramineae (*Poaceae*)

Description: Medium-height perennial grass, up to 4 feet tall, occurring singly or in clumps; rhizomes short; stems smooth; linear leaves (up to 12 inches long and ⅗ inch wide) with one main vein, both upper and lower surfaces smooth, margins and overlapping leaf sheaths rough; inconspicuous rough flowers borne in two rows on one side of elongate, fingerlike, ascending terminal spikes (2½–4¾ inches long), seven or more spikes present; yellowish seeds (grains).

Flowering period: June through October.

Habitat: Brackish marshes and sandy edges of tidal fresh marshes; nontidal swamps, wet prairies, moist to wet sandy areas, and edges of low woods.

Wetland indicator status: FACW.

Range: Southeastern North Carolina along the Coastal Plain to Florida, west to Alabama.

Salt Grass

Marsh Finger Grass

Bearded Sprangletop

Leptochloa fascicularis (Lam.) Gray

Grass Family

Gramineae (*Poaceae*)

Description: Medium-height annual grass, up to 3½ feet tall, occurring in clumps; stems smooth; linear, often inwardly rolled leaves (up to 10 inches long and about ⅕ inch wide) with rough surfaces and margins, smooth leaf sheaths; inconspicuous flowers borne in open terminal inflorescence (panicle, up to 12 inches long and 4 inches wide) with strongly ascending branches (usually not completely grown out of sheaths), six to twelve flowered, overlapping spikelets (less than ½ inch long) on rough, short stalks; yellowish red flat seed (grain).

Flowering period: Late summer.

Habitat: Salt, brackish, and tidal fresh marshes; nontidal marshes, pond and lake shores, wet sands, and disturbed areas.

Wetland indicator status: FACW+.

Range: New Hampshire south along the Coastal Plain to Florida, west to Texas; inland in Midwest and West; also in tropical America.

Similar species: Red Sprangletop (*L. filiformis*) has hairy leaf sheaths, red- to purple-tinged leaves and spikelets, a more open spreading panicle, and spikelets that are not overlapping or only barely so; it is FACW.

Key Grass

Monanthochloe littoralis Engelm.

Grass Family

Graminae (*Poaceae*)

Description: Low-growing, creeping perennial grass, up to 10 inches tall, forming extensive colonies; rhizomes and stolons hard; stems stiff, wiry, erect, and in clumps; short, ascending, wiry, grayish green leaves (mostly ⅕–½ inch long) densely overlapping in clusters along stem, internodes short or long; inconspicuous male and female flowers borne in the leaves on separate plants (dioecious).

Flowering period: Spring.

Habitat: Sand flats of irregularly flooded salt marshes, intertidal flats, and muddy estuarine shores.

Wetland indicator status: OBL.

Range: Peninsular Florida west to Louisiana and Texas; also in southern California, Mexico, and Cuba.

Torpedo Grass

Panicum repens L.

Grass Family

Graminae (*Poaceae*)

Description: Medium-height perennial grass, up to 2½ feet tall, often forming extensive colonies; rhizomes creeping and long, surface runners sometimes present; stems smooth with sheaths along lower part; simple, entire, flat or folded leaves (up to about 5 inches long and to ⅕ inch wide); inconspicuous flowers borne in open terminal inflorescence (panicle, up to 5 inches long) with strongly ascending branches.

Habitat: Tidal flats, estuarine beaches and shores; interdunal swales, shores of lakes and ponds, ditches, canals, and lower sand dunes.

Wetland indicator status: FACW−.

Range: Florida to Texas, mostly along the Coastal Plain; also in tropical America.

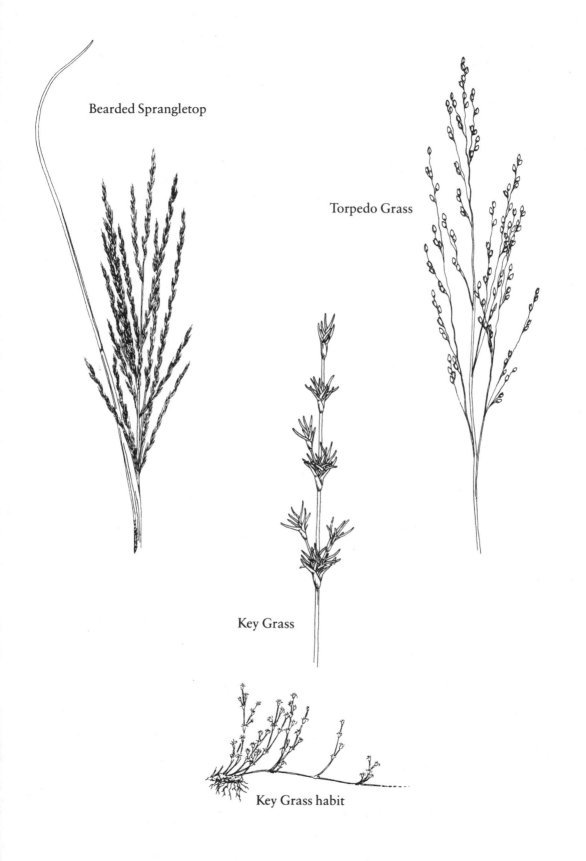

Bearded Sprangletop

Torpedo Grass

Key Grass

Key Grass habit

Switchgrass

Panicum virgatum L.

Grass Family
Gramineae (*Poaceae*)

Description: Medium-height to tall, erect perennial grass, up to 6½ feet high, forming dense clumps; hard, scaly rhizomes; stems stout, round, and erect; smooth, long, tapered leaves (up to 20 inches long and ⅘ inch wide), sometimes with few hairs at base; open terminal inflorescence (panicle, 8–16 inches long) many-branched and pyramid-shaped, branches fairly open with many spikelets on slender stalks.

Flowering period: June to October.

Habitat: Upper edges of salt marshes and irregularly flooded brackish and tidal fresh marshes; open woods, prairies, dunes, and shores.

Wetland indicator status: FAC+.

Range: Nova Scotia and Quebec to Manitoba and Montana, south to Florida, Texas, and Arizona.

Similar species: Fall Panic Grass (*P. dichotomiflorum*) is an annual grass lacking rhizomes; its panicles are terminal and axillary from sheaths of leafy branches, the latter more compressed since they are not completely grown out of the sheaths; it is found in slightly brackish and tidal fresh marshes; it is FACW. Beach Panic Grass (*P. amarum*, formerly *P. amarulum*) is a perennial grass with rhizomes usually rooting at nodes; its terminal panicle is compressed; it occurs on sand dunes, overwash sands, and interdunal swales; it is FAC.

Knotgrass or Joint Paspalum

Paspalum distichum L.

Grass Family
Gramineae (*Poaceae*)

Description: Medium-height, creeping perennial grass, up to 20 inches tall, often forming extensive mats; stolons rooting at nodes and nodes often hairy; linear leaves (up to 5 inches long and ⅓ inch wide) often folded (creased) and tapering to an inwardly rolled tip, leaf sheaths usually loose and long-hairy at top; inconspicuous flowers borne in terminal inflorescence composed of two (sometimes one or three) spreading or ascending spikes (racemes, up to 3⅓ inches long) with egg-shaped, fine-pointed, somewhat flattened spikelets borne in two rows along spike.

Flowering period: May through August.

Habitat: Brackish and tidal fresh marshes, sandy edges of salt marshes; nontidal marshes, ponds, and ditches.

Wetland indicator status: OBL.

Range: New Jersey south to Florida and Texas; also inland in Tennessee, Arkansas, and Idaho, and on the West Coast.

Similar species: Seashore Paspalum (*P. vaginatum*) occurs in similar habitats and is quite similar in appearance, but the second bract (glume) at the base of its spikelet is smooth, whereas that of Knotgrass is hairy; it is OBL. Some authors consider *P. vaginatum* to be the same as *P. distichum*. Vasey Grass (*P. urvillei*) has an inflorescence composed of eight to twenty-two spikes bearing long-haired spikelets; it is FAC. Longtom (*P. lividum*) has been reported in Texas salt marshes; it has a terminal inflorescence composed of three to eight curved spikes (up to 2 inches long); it is OBL. Bermuda Grass (*Cynodon dactylon*) has an inflorescence of usually four or five spikes with overlapping spikelets, and its ligule is a ring of whitish hairs; it is FACU. St. Augustine Grass (*Stenotaphrum secundatum*) also occurs in brackish marshes; it is a creeping, mat-forming grass with erect flowering stems (up to 12 inches tall), somewhat succulent leaves, and elongated spikes (about 4 inches long) bearing numerous spikelets singly or in pairs embedded in the joints of the somewhat wavy, thick flower stalk; it is FAC.

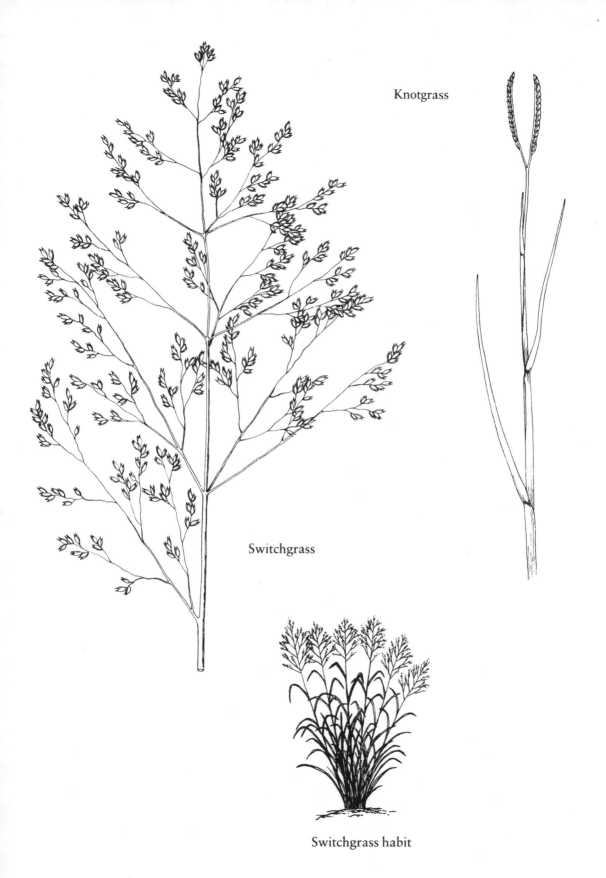

Knotgrass

Switchgrass

Switchgrass habit

Common Reed or Roseau Cane

Phragmites australis (Cav.) Trin. ex Steud.
[*Phragmites communis* Trin.]

Grass Family
Gramineae (*Poaceae*)

Description: Tall, erect perennial grass, 6½–
14 feet high, usually forming dense stands; rhizomes stout, sometimes creeping on surface; stems round, hollow, and thick; long, flat, tapering (long-pointed) leaves (up to 24 inches long and 2 inches wide) distinctly arranged in two ranks; somewhat open and drooping to dense and erect, many-branched terminal inflorescence (panicle, 8–16 inches long) with silky, light brown hairs beneath on stem, branches usually somewhat drooping; flower clusters usually purplish when young and white or light brown and feathery when mature.

Flowering period: Late July to October.

Habitat: Brackish and tidal fresh marshes (regularly and irregularly flooded zones), also upper edges of salt marshes and old spoil deposits; nontidal marshes, swamps, wet shores, seepage areas, ditches, spoil embankments, and disturbed areas.

Wetland indicator status: FACW.

Range: Nova Scotia and Quebec to British Columbia, south to Florida, Texas, and California.

Similar species: See Giant Plume Grass (*Erianthus giganteus*).

dense inflorescence

open inflorescence

Common Reed

Foxtail Grass

Setaria geniculata (Lam.) Beauv.

Grass Family
Gramineae (*Poaceae*)

Description: Medium-height, erect perennial grass, up to 2½ feet tall; short, knotty rhizomes; stems round, hollow, and erect or sometimes lying flat on ground at base, then ascending; long, tapering, mostly flat leaves (up to 8 inches long and ¼ inch wide); dense terminal spikelike inflorescence (panicle, 1–4 inches long and less than ½ inch wide) with many spikelets and light brown bristles (four or more below each spikelet).

Flowering period: May to October.

Habitat: Brackish marshes and upper edges of salt marshes; moist to dry ground and waste places.

Wetland indicator status: FAC.

Range: Massachusetts south to Florida and Texas; inland in the North to Pennsylvania, Illinois, Kansas, New Mexico, and California.

Similar species: Yellow-bristle Foxtail (*S. glauca*) is an annual lacking rhizomes, and its leaf sheaths are folded (keeled); it is FAC. Rabbitfoot Grass (*Polypogon monspeliensis*) is also an annual lacking rhizomes, but its inflorescence is longer (up to 6 inches long), wider (up to ⅘ inch wide), bristly hairy, and yellowish, and its leaves are rough; it is FACW. Giant Foxtail Grass (*S. magna*) grows in brackish marshes from New Jersey south; it is up to 15 feet tall and has one to three bristles below each spikelet; it is FACW+.

Giant Foxtail Grass

Setaria magna Griseb.

Grass Family
Gramineae (*Poaceae*)

Description: Tall annual grass, up to 15 feet high; stems with thick bases and somewhat soft, hairy internodes; leaves (up to 24 inches long and to 2 inches wide) with rough surfaces and margins, smooth leaf sheaths; dense, cylinder-shaped terminal inflorescence (spike, 4–20 inches long and 1–2 inches wide) with long-hairy, angled main stem (rachis), two-flowered spikelets with one to three long bristles (up to 1⅕ inches long, much longer than spikelet) at base.

Flowering period: August through October.

Habitat: Brackish and tidal fresh marshes; nontidal marshes, swamps, and bottomlands.

Wetland indicator status: FACW+.

Range: New Jersey south to Florida, west to Texas; also in the West Indies.

Similar species: Foxtail Grass (*S. geniculata*) is shorter, to 2½ feet tall, and has four to twelve long bristles at base of each spikelet; it is FAC.

Foxtail Grass

Giant Foxtail Grass

Smooth Cordgrass, Saltwater Cordgrass, or Oyster Grass

Spartina alterniflora Loiseleur

Grass Family
Gramineae (Poaceae)

Description: Low to tall, erect perennial grass, 1–8 feet high; stems stout, round, and hollow, often soft and spongy at base; elongate, smooth leaves (up to 16 inches long and ½ inch wide) tapering to a long point with inwardly rolled tip, leaf margins smooth or weakly rough, sheath margins hairy; narrow terminal inflorescence (panicle, usually 4–12 inches long) composed of five to thirty spikes (2–4 inches long) alternately arranged and appressed to main axis with ten to fifty sessile spikelets along one side of the axis of each spike. *Note:* Salt crystals can often be seen on its leaves during the growing season; two major growth forms are generally recognized— (1) short form (less than 1½ feet, often having yellowish green leaves, and characteristic of irregularly flooded high marsh), and (2) tall form (greater than 1½ feet and typical of regularly flooded low marsh); some ecologists also recognize an intermediate growth form (1½–3 feet tall).

Flowering period: June through October.

Habitat: Salt and brackish marshes (regularly and irregularly flooded zones).

Wetland indicator status: OBL.

Range: Quebec and Newfoundland to Florida and Texas.

Big Cordgrass or Hogcane

Spartina cynosuroides (L.) Roth

Grass Family
Gramineae (Poaceae)

Description: Tall, erect perennial grass, up to 10 feet high; stems stout, round, and hollow; elongate leaves (up to 28 inches long and 1 inch wide) tapering to a point, margins very rough; open terminal inflorescence (panicle, 4–12 inches long) composed of twenty to fifty erect, crowded spikes (1¼–2⅗ inches long, uppermost spikes usually shorter than lower ones), each with many (up to seventy) densely overlapping spikelets borne on one side of the axis.

Flowering period: June into October.

Habitat: Irregularly flooded salt, brackish, and tidal fresh marshes.

Wetland indicator status: OBL.

Range: Massachusetts to Florida and Texas.

Smooth Cordgrass habit

Overlapping leaf sheaths of lower leaves

Smooth Cordgrass

Big Cordgrass

Salt Meadow Cordgrass, Marsh-hay Cordgrass, or Wiregrass

Spartina patens (Ait.) Muhl.

Grass Family
Gramineae (Poaceae)

Description: Low to medium-height, erect or spreading perennial grass, usually 1–3 feet tall, often forming cowlicked mats; stems slender (wirelike), stiff, and hollow; very narrow linear leaves (less than ⅕ inch wide and up to 1½ feet long) with margins rolled inwardly; open terminal inflorescence (panicle, up to 8 inches long) usually composed of three to six spikes (⅘–2 inches long), alternately arranged and diverging from main axis at 45–60-degree angles, each with twenty to fifty densely overlapping spikelets (⅕–½ inch long) borne on one side of the axis.

Flowering period: June into October.

Habitat: Irregularly flooded salt, brackish, and tidal fresh marshes (often forming cowlicked mats, and reported to occur at times in regularly flooded zone); on wet beaches, sand dunes, and borders of salt marshes (var. *monogyna*); also inland saline areas.

Wetland indicator status: FACW.

Range: Quebec to Florida and Texas; inland in New York and Michigan.

Similar species: Other *Spartina* members have stout stems, not wirelike. *S. patens* grows in two forms: The typical form lies flat (decumbent) with upward-spreading stems that create the "cowlicks" of the high salt marsh, whereas the variety *monogyna* grows upright and straight and colonizes drier sites. Sand Cordgrass (*S. bakerii*) has a somewhat similar inflorescence, but it is more appressed, grows in dense clumps, and blooms mostly in winter and spring; it is FACW+.

Gulf Cordgrass or Marsh Bunchgrass

Spartina spartinae (Trin.) Merrill ex. A. Hitchc.

Grass Family
Gramineae (Poaceae)

Description: Medium-height to tall perennial grass, usually up to 4 feet high (rarely to 6½ feet), growing in dense clumps; rhizomes absent; stems numerous, smooth, and unbranched; elongate, inwardly rolled leaves (up to 28 inches long and to ⅕ inch wide) with spinelike tips, margins somewhat rough; inconspicuous flowers borne in ten or more dense, overlapping spikes forming a cylinder-shaped, spikelike terminal inflorescence (panicle, up to 16 inches long and less than ⅓ inch wide) usually tapered at both ends.

Flowering period: March into October.

Habitat: Irregularly flooded salt and brackish marshes and salt flats; coastal wet prairies, nontidal marshes, and swamps.

Wetland indicator status: OBL.

Range: Southern Florida to Texas; also in Mexico and Central America.

Similar species: Sand Cordgrass (*S. bakerii*) grows in dense clumps and usually lacks rhizomes, but its spikes are somewhat spreading, not closely overlapping; it occurs in sandy margins of salt and brackish marshes, coastal swales, wet prairies, and wet pinelands from South Carolina to Florida; it is FACW+.

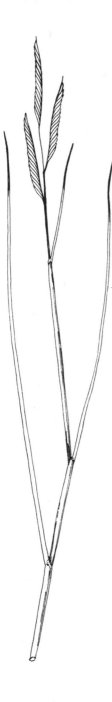

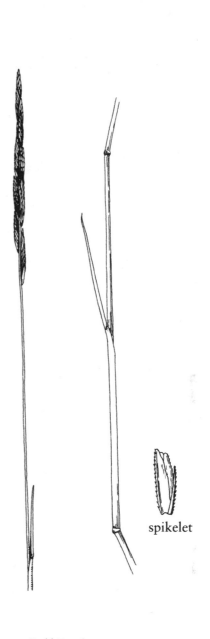

spikelet

Gulf Cordgrass

Salt Meadow Cordgrass

Sand Cordgrass spikelet

Coastal Dropseed

Sporobolus virginicus (L.) Kunth

Grass Family
Gramineae (Poaceae)

Description: Low-growing, erect, creeping perennial grass, up to 18 inches tall; rhizomes wiry; linear leaves (up to 5 inches long and less than ⅕ inch wide) rolled inwardly and internally divided into partitions by cell walls (septate), ligule hairy, upper leaves conspicuously shorter than lower leaves, leaf sheaths overlapping; one-flowered spikelets (less than ⅕ inch long) borne in narrow, dense terminal inflorescence (panicle, 1⅕–4 inches long and less than ⅗ inch wide) with rough branches; reddish seeds.

Flowering period: September through October.

Habitat: Irregularly flooded salt marshes, salt barrens, and sandy or muddy saline shores.

Wetland indicator status: FACW+.

Range: Southeastern Virginia to Florida, west to Texas; also in Mexico, tropical America, and the West Indies.

Similar species: Spike Grass or Salt Grass (*Distichlis spicata*) has a wider, denser panicle (to ⅘ inch wide) with many-flowered (three to twelve) spikelets that are larger (2–6 inches long) and its uppermost leaves are not greatly reduced in size and may often exceed panicle; it is FACW+.

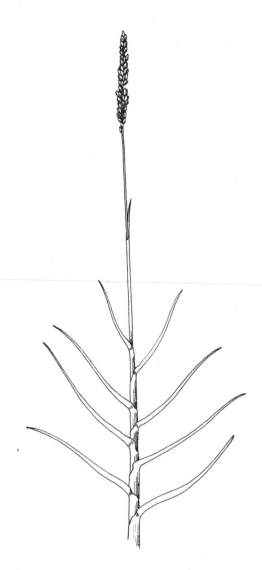

Coastal Dropseed

SEE ALSO Lowland Broomsedge (*Andropogon glomeratus*) and Walter Millet (*Echinochloa walteri*).

Nuttall's Cyperus

Cyperus filicinus Vahl

Sedge Family
Cyperaceae

Description: Low to medium-height, erect, grasslike annual herb, 4–16 inches tall; stems three-angled; elongate, linear leaves (usually less than $1/10$ inch wide); inconspicuous flowers borne in sessile or stalked spikes (up to 5 inches long) forming clusters, several clusters forming terminal inflorescence (umbel) subtended by several leafy bracts (up to 10 inches long), flowers covered by green (immature) or straw-colored (mature) sharp-pointed scales with three to five nerves clustered together at center; flattened, lens-shaped, brownish nutlet (achene, about $1/20$ inch long).

Flowering period: August into October.

Habitat: Brackish marshes (regularly and irregularly flooded zones); sandy coastal beaches and, rarely, inland shores of ponds.

Wetland indicator status: OBL.

Range: Southern Maine to Florida and Louisiana.

Similar species: Many-spike Flatsedge (*C. polystachyos*) is an annual with a stem having reddish and somewhat fibrous bases and reddish or brownish lens-shaped nutlets (less than $1/20$ inch long); it is FACW. Slender Cyperus (*C. filiculmis*) is a FACU+ perennial plant with bulblike rhizomes, somewhat rough-margined leaves, and scales with seven to thirteen well-spaced nerves. Other *Cyperus* spp. resembling *C. filicinus* are found in tidal fresh marshes. Yellow Cyperus (*C. flavescens*), an OBL species, has yellowish green scales about half as wide as long, whereas the scales of *C. filicinus* are roughly three times as long as wide. Shining Cyperus (*C. rivularis*) has reddish brown, blunt-tipped scales at maturity; it is FACW. See Fragrant Galingale (*C. odoratus*), Retrorse Flatsedge (*C. retrorsus*), and Sheathed Flatsedge (*C. haspan*).

scale

Nuttall's Cyperus

Gulf Coast Spike-rush

Eleocharis cellulosa Torr.

Sedge Family
Cyperaceae

Description: Medium-height, erect, perennial grasslike plant, 6–40 inches tall; roundish, unbranched stems with leaves reduced to basal leaf sheaths with fine-pointed tips; many inconspicuous flowers borne in a single, cylinder-shaped terminal spikelet (to 2 inches long) that is not wider than the stem, flower scales straw-colored with rough margins; olive or brown nutlet (achene) with netted pattern and a triangle-shaped projection (tubercle) at one end and usually six bristles attached at other end.

Flowering period: July through September.

Habitat: Brackish and tidal fresh marshes; nontidal marshes (usually in shallow water).

Wetland indicator status: OBL.

Range: North Carolina south to Florida, west to Texas; also in Mexico, Bermuda, the West Indies, and Central America.

Similar species: Horsetail Spike-rush (*E. equisetoides*) is quite similar, but its round stem is divided into partitions (septate); it is OBL. Square-stemmed Spike-rush (*E. quadrangulata*) also has a terminal spikelet that is not much wider than the stem; its stems are triangular or squarish in cross-section; it occurs in freshwater marshes from Massachusetts south; it is OBL. Other spike-rushes have terminal spikelets that are budlike, much thicker than the stems.

Dwarf Spike-rush

Eleocharis parvula (Roem. & J. A. Schultes) Link ex Bluff & Fingerh.

Sedge Family
Cyperaceae

Description: Low-growing, erect, grasslike herb, usually less than 3 inches but up to 5 inches tall, forming mats; stems spongy and threadlike with no apparent leaves, leaves actually reduced to stem sheaths; inconspicuous flowers covered by green, straw-colored, or brown scales borne on a single terminal budlike spikelet (much wider than stem); three-angled nutlet (achene).

Flowering period: July to October.

Habitat: Salt and brackish marshes (regularly and irregularly flooded zones) and tidal fresh marshes (less commonly); wet inland saline soils.

Wetland indicator status: OBL.

Range: Newfoundland to Florida and Texas; inland locally in western New York, Michigan, and Missouri; also on Pacific coast.

Similar species: Pale Spike-rush (*E. flavescens*) and White Spike-rush (*E. albida*) are mat-forming but taller; the former often has spongy stems, whereas the latter has wiry stems; both are OBL. Positive identification requires examining the nutlets (achenes) and reference to a taxonomic manual.

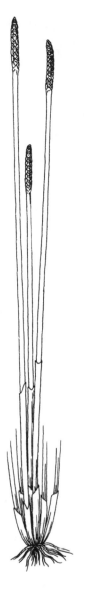

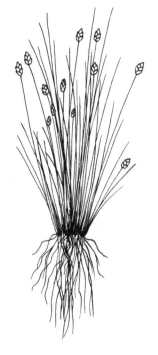

Gulf Coast Spike-rush   Dwarf Spike-rush

Beaked Spike-rush

Eleocharis rostellata (Torr.) Torr.

Sedge Family
Cyperaceae

Description: Medium-height to tall, erect, perennial grasslike plant, 1–3½ feet high, forming dense clumps; stems elongate and flattened with no apparent leaves, leaves actually reduced to stem sheaths; inconspicuous flowers covered by brownish scales borne on a single terminal budlike spikelet (much wider than stem); three-angled nutlet (achene).

Flowering period: July into October.

Habitat: Irregularly flooded salt, brackish, and tidal fresh marshes; calcareous nontidal marshes and swamps.

Wetland indicator status: OBL.

Range: Nova Scotia to Florida, along the coast; inland locally from New York to Wisconsin, south to Ohio and Indiana; also in western states.

Similar species: See Long-tubercle Spike-rush (*E. tuberculosa*). Other spike-rushes found in coastal marshes besides those described in this book include White Spike-rush (*E. albida*; OBL), Purple Spike-rush (*E. atropurpurea*; FACW), Capitate Spike-rush (*E. caribaea*; FACW), Engelmann's Spike-rush (*E. engelmannii*; FACW), Creeping Spike-rush (*E. fallax*; OBL), Pale Spike-rush (*E. flavescens*; OBL), Clustered Spike-rush (*E. geniculata*; FACW+), Salt Marsh Spike-rush (*E. halophila*; OBL), Matted Spike-rush (*E. intermedia*; FACW), Sand Spike-rush (*E. montevidensis*; FACW+), Blunt Spike-rush (*E. obtusa*; OBL), and Three-angle Spike-rush (*E. tricostata*; FACW+). Identification requires examination of nutlets and reference to a taxonomic manual.

Salt Marsh Fimbristylis

Fimbristylis castanea (Michx.) Vahl
[*Fimbristylis spadicea* (L.) Vahl]

Sedge Family
Cyperaceae

Description: Low to medium-height, erect, grasslike perennial herb, 8–40 inches tall; stems slender, triangular in cross-section, stiff, somewhat enlarged at base; elongate linear leaves (shorter than stem) tapering to a long point, rolled inwardly, arising from near base of plant; inconspicuous flowers covered by dark, glossy brown scales in budlike spikelets (⅕–1 inch long and about ½ inch wide) borne on slender stalks and forming terminal inflorescence, surrounded and often overtopped by two or three leaflike bracts; dull brown, lens-shaped nutlet (achene).

Flowering period: July to October.

Habitat: Irregularly flooded salt and brackish marshes; interdunal swales, and moist coastal sands and marls.

Wetland indicator status: OBL.

Range: Long Island to Florida and Texas.

Similar species: Carolina Fimbristylis (*F. caroliniana*) also occurs in salt marshes in the same range and is FACW+; its basal sheaths are shorter and thinner, and its bracts are shorter; its scales are short-hairy with the midvein forming an elevated keel toward the tip, whereas *F. castanea* has smooth scales without a keel. Slender Fimbristylis (*F. autumnalis*) is an annual, to 16 inches tall, having a many-branched terminal inflorescence bearing many slender spikelets (⅕ inch or less long and about ¹⁄₂₅ inch wide) with three-sided nutlets; it is OBL. Hurricane-grass (*F. spathacea*) occurs on salt flats and along the edges of salt and brackish marshes in southern Florida; its spikelets are arranged in dense headlike clusters; it is FACW+.

Beaked Spike-rush   Salt Marsh Fimbristylis

Olney's Three-square

Scirpus americanus Pers.
[*Scirpus olneyi* Gray]

Sedge Family
Cyperaceae

Description: Medium-height to tall, erect, perennial grasslike herb, up to 7 feet high; long, hard rhizomes; stems stout and sharply triangular (in cross-section) with deeply concave sides; no apparent leaves; inconspicuous flowers borne in five to twelve sessile budlike spikelets covered by brown scales located very near and almost at top of pointed stem (portion above spikelets is ½–2 inches long); dark gray to black nutlet (achene).

Flowering period: July into September.

Habitat: Irregularly flooded brackish marshes, upper edges of salt marshes, and tidal fresh marshes (less commonly); inland saline areas.

Wetland indicator status: OBL.

Range: New Hampshire and western Nova Scotia to Florida, Texas, and Mexico; inland in New York, Michigan, and western states; also along the Pacific coast.

Similar species: The stem of Common Three-square (*S. pungens*, formerly *S. americanus*) is stout and triangular but not deeply concave; its stem is also occasionally twisted; it is OBL. Also, its spikelets are not usually located as close to the top of the stem as they are in *S. americanus*. Other plants with stout, triangular stems have leafy bracts at end of stem, such as umbrella sedges or flatsedges (*Cyperus* spp.).

Salt Marsh Bulrush

Scirpus robustus Pursh

Sedge Family
Cyperaceae

Description: Medium-height, erect, grasslike perennial herb, up to 5 feet tall; thick rhizome; stems stout and triangular; several elongate, linear, grasslike leaves (½ inch wide) tapering to a long point; inconspicuous flowers borne in three or more budlike spikelets (mostly sessile, few stalked) covered by brown scales, inflorescence surrounded by two to four elongate, erect or somewhat erect, leaflike bracts; dark brown to black nutlet (achene).

Flowering period: July to October.

Habitat: Irregularly flooded salt and brackish marshes (occasionally regularly flooded zones).

Wetland indicator status: OBL.

Range: Nova Scotia to Florida and Texas.

Similar species: Canby's Bulrush (*S. etuberculatus*) has one leafy bract (2–10 inches long) subtending an inflorescence of several elongate cylinder-shaped spikelets (mostly stalked) at the top of the stem, reddish rhizomes, and elongate triangular (in cross-section) leaves (up to 32 inches long); it is OBL and occurs in brackish and tidal fresh marshes from Delaware south to Florida and Texas.

Olney's Three-square   Salt Marsh Bulrush

RUSHES

Black Grass

Juncus gerardii Loiseleur
Rush Family
Juncaceae

Description: Low to medium-height, erect, grasslike perennial herb, 8–24 inches tall; one or two elongate linear leaves (up to 8 inches long) round in cross-section, uppermost located near middle of stem; flowers borne on erect or somewhat erect, branched inflorescence (1–3¼ inches long); dark brown fruit capsule.

Flowering period: June into September.

Habitat: Irregularly flooded salt marshes (usually at upper elevations and sometimes forming cowlicked mats) and occasionally, brackish marshes.

Wetland indicator status: OBL.

Range: Quebec and Newfoundland to Virginia, reported to Florida; also inland in New York, Indiana, and Minnesota; also in Pacific states.

Black Needlerush or Black Rush

Juncus roemerianus Scheele
Rush Family
Juncaceae

Description: Medium-height to tall, sharp-pointed, evergreen grasslike plant, up to 6½ feet tall; reddish lower stem and rhizomes; stiff, very sharp-pointed leaves, round in cross-section, olive-brown to grayish green in color, appearing as unbranched linear stems; inconspicuous greenish or light brown flowers borne in clusters of two to eight, appearing laterally above middle of stem; reddish brown, three-sided fruit capsules (⅓ inch long or less) bearing finely ribbed seeds. (*Note:* Grayish colored dead "stems" may be present.)

Flowering period: March through October.

Habitat: Brackish marshes, upper edges of salt marshes, and edges of salt barrens.

Wetland indicator status: OBL.

Range: Southernmost Delaware and Maryland south to Florida, west to southernmost Texas.

Similar species: See Canada Rush (*J. canadensis*), Soft Rush (*J. effusus*), and Needle-pod Rush (*J. scirpoides*).

Black Needlerush

Black Grass

OTHER GRASSLIKE PLANTS

Narrow-leaved Cattail
Typha angustifolia L.
Cattail Family
Typhaceae

Description: Medium-height to tall, erect, perennial herb, up to 6 feet high; simple, entire, elongate, linear basal leaves (⅕–½ inch wide), flattened (plano-convex), sheathing at base and ascending along stem in an apparent alternately arranged fashion, usually less than ten leaves; inconspicuous flowers borne on long stalk and arranged in two terminal cylinder-shaped spikes (male spike above female spike) separated by a space, female spike green in spring and brown in summer at maturity and persistent in winter, male spike covered with yellow pollen grains at maturity and then disintegrating (nonpersistent).

Flowering period: Late May through July.

Habitat: Brackish and tidal fresh marshes (regularly and irregularly flooded zones); nontidal fresh and alkaline marshes.

Wetland indicator status: OBL.

Range: Nova Scotia, Quebec, and Ontario south to Florida and Texas, especially abundant along the coast.

Similar species: Broad-leaved Cattail (*T. latifolia*) grows taller (to 10 feet) and has wider leaves (to 1 inch) and no space between male and female spikes. Southern Cattail (*T. domingensis*) occurs from Delaware and Maryland south; it is much taller (8–13 feet) with ten or more leaves and a space between male and female spikes. Blue Cattail (*T. glauca*) is a hybrid between *T. latifolia* and *T. angustifolia*; it has a yellowish buff-colored pith, whereas the other two have a white pith. All cattails are OBL.

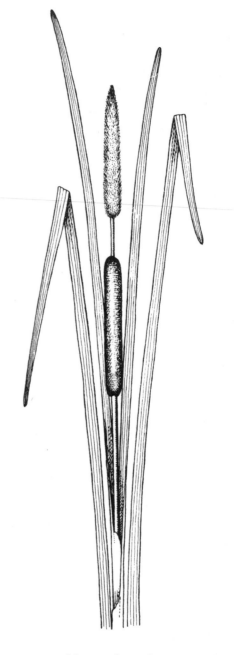

Narrow-leaved Cattail

FLESHY FLOWERING HERBS

Sea Purslane

Sesuvium portulacastrum (L.) L.

Carpet-weed Family
Aizoaceae

Description: Fleshy, creeping, mat-forming perennial herb; smooth stems rooting at nodes; simple, entire, fleshy, linear to spoon-shaped leaves (up to 2 inches long and to ¾ inch wide) with somewhat pointed tips and wedge-shaped bases, oppositely arranged; small pinkish to light purplish five-"petaled" flowers (less than 1 inch wide) with a prominent tip (horn) at the end of the "petals," borne singly on stalks from leaf axils; fruit capsule (about ½ inch long) bearing shiny black seeds.

Flowering period: May into November.

Habitat: Irregularly flooded sand flats or barrens, mangrove swamps, and sandy borders of salt marshes; interdunal swales, sand dunes, upper sandy beaches, and spoil areas.

Wetland indicator status: FACW.

Range: North Carolina to Florida, west to Texas; also in tropical America.

Similar species: Sea Purslane (*S. maritimum*) is an annual plant that does not root at the nodes; its leaves have rounded tips, and its flowers and fruits are short-stalked or not stalked (sessile); it occurs in similar habitats from Long Island, New York, south; it is FACW. Common Purslane (*Portulaca oleracea*) and Pink Purslane (*P. pilosa*) are succulent herbs with erect branches; the former species has purplish red stems and branches and yellow four- or five-"petaled" flowers borne singly or in terminal clusters, whereas the latter has alternate leaves with many long hairs in the axils; both are FACU.

Sea Purslane

Silverhead or Marsh Samphire

Philoxerus vermicularis (L.) R. Br.
    ex J. E. Smith

Amaranth Family
Amaranthaceae

Description: Somewhat fleshy, perennial, mat-forming herb; stems creeping with somewhat erect branches (up to 20 inches tall); simple, entire, fleshy, linear to oblong leaves (up to 2½ inches long and to ½ inch wide) tapering at base, hairy leaf axils, oppositely arranged; many minute, five-"petaled" white flowers borne in terminal and axillary headlike inflorescences (about ½ inch long and less than ½ inch wide); small brownish seeds.

Flowering period: Summer and fall.

Habitat: Sandy areas of irregularly flooded salt marshes and mangrove swamps; sand dunes, beaches, and wet marls.

Wetland indicator status: FACW+.

Range: Florida to Texas and Mexico; also in the Caribbean and tropical America.

Seaside Heliotrope

Heliotropium curassavicum L.

Borage Family
Boraginaceae

Description: Short to medium-height, erect, annual or short-lived perennial succulent herb, up to 2 feet tall; stems succulent; simple, entire, nearly veinless, linear lance-shaped leaves (up to 2½ inches long) with blunt or rounded tips, short-stalked or stalkless, lower leaves reduced, alternately arranged; many small, whitish (sometimes blue- or purple-tinged), five-lobed tubular flowers (less than ⅕ inch long) borne on terminal or axillary inflorescences (racemes, up to 5 inches long) that curl at their tips before flowering; small fruit capsule divided into four nutlets (one-seeded).

Flowering period: June through September.

Habitat: Regularly flooded and irregularly flooded salt and brackish marshes, edges of mangrove swamps, and sandy areas along the upper borders of salt marshes.

Wetland indicator status: OBL.

Range: Massachusetts south to Florida, west to Oklahoma and Texas; also in Arizona, New Mexico, Mexico, and tropical America.

Similar species: Other heliotropes do not have succulent leaves. Pineland Heliotrope (*H. polyphyllum*) occurs along brackish shores in Florida; it is a FAC perennial covered with stiff hairs and with sessile leaves and flowers subtended by leaflike bracts. Taper-leaf Heliotrope (*H. angiospermum*) occurs on shell mounds and edges of estuarine shores and mangrove swamps in South Florida; it is a FACU annual with short-stalked leaves (at least the lower leaves).

Silverhead

Seaside Heliotrope

Salt Marsh Sand Spurrey

Spergularia marina (L.) Griseb.

Pink Family
Caryophyllaceae

Description: Low-growing, erect or nearly creeping, fleshy annual herb, up to fourteen inches long; stem simple or much branched, smooth or finely hairy, bearing glands; simple, entire, linear fleshy leaves (⅕–1⅗ inches long) with triangular structures (stipules) at leaf bases, oppositely arranged; small pink or white five-petaled flowers (⅙ inch wide) borne on stalks from upper leaf axils.

Flowering period: April through October.

Habitat: Irregularly flooded salt and brackish marshes (usually in sandy pannes); inland alkaline areas.

Wetland indicator status: OBL.

Range: Quebec to British Columbia, south to Florida and southern California; inland locally to Illinois, Texas, and New Mexico.

Similar species: Sea Blite (*Suaeda linearis*) and related species have alternately arranged, fleshy, linear leaves and small green flowers borne on terminal spikes; Sea Blite is OBL. Dwarf Pearlwort (*Sagina decumbens*) has been reported in coastal marshes; it is not succulent, lacks the triangular stipule, and grows to about 6 inches tall; it is FAC+.

Marsh Orach or Spearscale

Atriplex patula L.

Goosefoot Family
Chenopodiaceae

Description: Low to medium-height, erect or prostrate, fleshy annual herb, up to 3½ feet tall or long; stems grooved; simple, entire, arrow-head-shaped, sometimes narrow or lance-shaped, light green (with whitish bloom) fleshy leaves (up to 3 inches long) on stalks (petioles), mostly alternately arranged and sometimes oppositely arranged, especially lower leaves; very small green flowers borne in somewhat ball-shaped clusters on open, nearly leafless spikes at upper leaf nodes.

Flowering period: July to November.

Habitat: Irregularly flooded salt and brackish marshes; inland saline or alkaline soils, edges of sand dunes, and waste places.

Wetland indicator status: FACW.

Range: Prince Edward Island and Nova Scotia to British Columbia, south to South Carolina, Missouri, and California.

Similar species: Seabeach Orach (*A. arenaria*) is grayish green in color, and its leaves are somewhat lance-shaped; it commonly occurs on sand dunes and may be found in the upper high salt marsh; it is FAC.

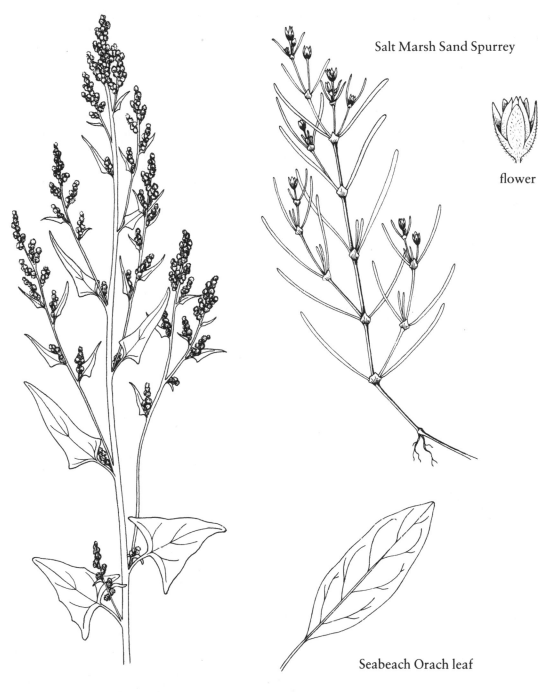

Salt Marsh Sand Spurrey

flower

Marsh Orach

Seabeach Orach leaf

Lamb's-quarters or Pigweed

Chenopodium album L.

Goosefoot Family
Chenopodiaceae

Description: Medium-height to tall, fleshy-leaved annual herb, up to 10 feet high, usually much shorter in coastal marshes; stems unbranched below, much branched above, and grooved; simple, irregularly coarse-toothed, thick, fleshy, pale green, somewhat triangle-shaped leaves (to 3 inches long and to 2 inches wide), smooth above, whitish powdery below, upper leaves entire and lance-shaped, alternately arranged; inconspicuous greenish flowers borne in spikelike inflorescences (up to 12 inches long) from leaf axils or top of stem; bladderlike fruit capsule bearing one seed.

Flowering period: March into November.

Habitat: Salt and brackish marshes; inland salt flats, floodplains, waste places, cultivated fields, and barnyards.

Wetland indicator status: FAC−.

Range: Newfoundland to Alaska, south to Florida and Arizona; also in Mexico and tropical America; native of Eurasia.

Similar species: Mexican Tea (*C. ambrosioides*) is strongly aromatic (pungent), covered with resin dots, and has leafy inflorescences; it is FACU.

Common Glasswort

Salicornia europaea L.

Goosefoot Family
Chenopodiaceae

Description: Low-growing, erect, fleshy annual herb, 4–20 inches tall; stems fleshy, jointed, erect, and much branched or lower branches creeping; leaves reduced to minute scales, blunt or rounded tips (below spikes), oppositely arranged; inconspicuous green flowers in upper joints of stem forming spikes (less than ⅕ inch wide). (*Note:* Plant turns pinkish red to red in the fall.)

Flowering period: July to November.

Habitat: Irregularly flooded salt marshes (usually in sandy pannes) and salt flats; inland saline soils and marshes.

Wetland indicator status: OBL.

Range: Quebec and Newfoundland to Florida; inland in New Brunswick, New York, and Michigan; Alaska to California.

Similar species: Bigelow's Glasswort (*S. bigelovii*) has sharp-tipped scales below the spikes, thicker spikes (⅕–¼ inch wide), and does not have creeping lower branches; it turns yellowish orange in the fall; it is OBL.

Perennial Glasswort or Woody Glasswort

Salicornia virginica L.
[includes *Salicornia perennis* Mill.]

Goosefoot Family
Chenopodiaceae

Description: Low-growing, erect, fleshy perennial herb, up to 12 inches tall; main stem somewhat woody-cored, fleshy, and jointed, creeping to form mats, with erect flowering branches; leaves reduced to minute scales, oppositely arranged; inconspicuous green flowers in upper joints of stem.

Flowering period: July into October.

Habitat: Irregularly flooded salt marshes (usually in sandy pannes), salt flats, and mangrove swamps.

Wetland indicator status: OBL.

Range: Southern New Hampshire to Florida and Texas; also Alaska to California.

Similar species: Other glassworts (*Salicornia* spp.) are annuals with stems lacking woody cores; they are not mat-forming.

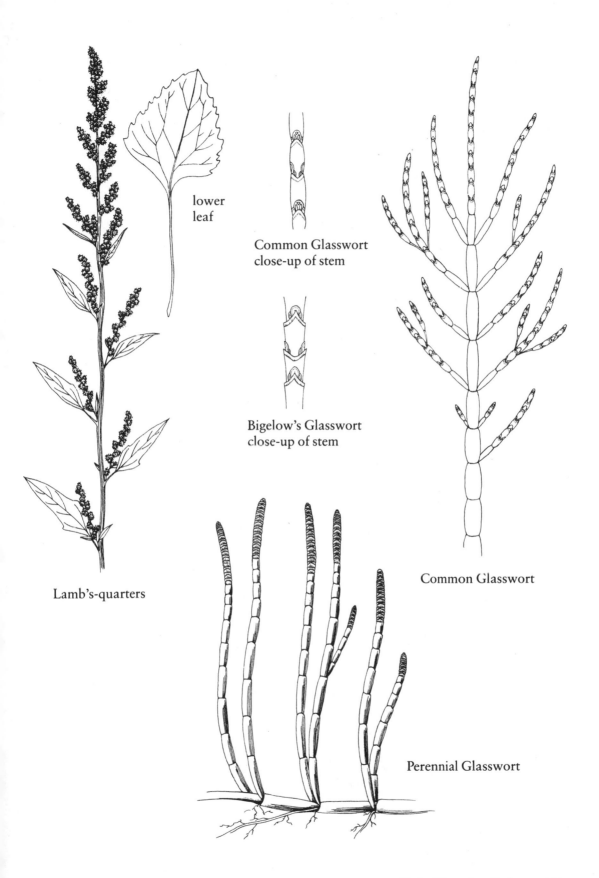

lower
leaf

Common Glasswort
close-up of stem

Bigelow's Glasswort
close-up of stem

Lamb's-quarters

Common Glasswort

Perennial Glasswort

Russian Thistle or Saltwort

Salsola kali L.

Goosefoot Family
Chenopodiaceae

Description: Low to medium-height, erect, fleshy and prickly annual herb, 1–3 feet tall; stems smooth or hairy, often marked with red or purplish vertical lines; simple, entire, fleshy, prickly leaves (up to 2 inches long), lower leaves somewhat cylinder-shaped, upper leaves shorter and stiff, with long-spined tip, alternately arranged; small green flowers borne singly or in twos or threes on short spike from axils of upper leaves.

Flowering period: June to October.

Habitat: Beaches and upper edges of irregularly flooded salt marshes, sometimes on top of tidal wrack (vegetation debris—leaves and stems), and also in pannes.

Wetland indicator status: FACU.

Range: Newfoundland along the Coastal Plain to Florida and Texas; also inland from Minnesota to Washington and California; native of Eurasia.

Sea Blite

Suaeda linearis (Elliott) Moq.

Goosefoot Family
Chenopodiaceae

Description: Low to medium-height, erect, fleshy annual herb, up to 32 inches tall; stems grooved, often red-tinged (late in season), and usually much branched; simple, entire, dark green, linear fleshy leaves (up to 2 inches long), usually flat on one side and rounded on the other (in cross-section), upper leaves reduced in size, alternately arranged; small green flowers borne on terminal spikes, either singly or in clusters of threes in upper leaf axils.

Flowering period: August to November.

Habitat: Irregularly flooded salt marshes (often in sandy pannes), salt flats, and mangrove swamps.

Wetland indicator status: OBL.

Range: Maine to Florida and Texas.

Similar species: A Eurasian introduction, Sea Blite (*S. maritima*) is quite similar to *S. linearis*, but has pale green, usually whitened, leaves; it occurs from Virginia north and has been reported in Florida; it is OBL. Shrubby Sea Blite (*S. conferta*) is a shrub with brittle branches and bluish gray, fleshy, oblong leaves; it grows in coastal marshes in Texas and eastern Mexico; it is OBL.

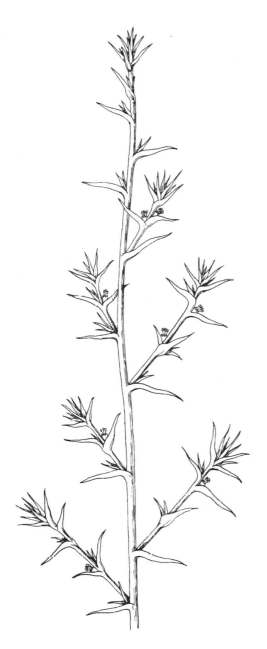

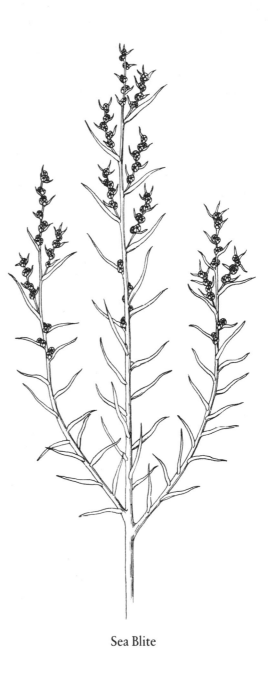

Russian Thistle

Sea Blite

Annual Salt Marsh Aster

Aster subulatus Michx.
[*Aster exilis* Elliott]
Composite or Aster Family
Compositae (Asteraceae)

Description: Low to medium-height, erect, somewhat fleshy annual herb, 4–32 inches tall (sometimes to 6 feet high), with a short taproot; simple, entire, somewhat fleshy, linear or narrowly lance-shaped leaves (up to 6 inches long), alternately arranged; small purplish or blue flowers in heads (less than ½ inch wide), usually with very short, almost inconspicuous rays, heads usually in an open inflorescence.

Flowering period: June through November.

Habitat: Irregularly flooded salt, brackish, and tidal fresh marshes, and tidal flats; nontidal marshes and thickets and edges of woods.

Wetland indicator status: OBL.

Range: New Brunswick, southern Maine, and New Hampshire, south to Florida and Louisiana; reported inland near Syracuse, New York, and Detroit, Michigan.

Similar species: Coastal Plain Aster (*A. racemosus*) has small, bluish to lavender daisylike flowers (less than ½ inch wide) with conspicuous rays and fine-toothed, nonfleshy leaves; it is FACW.

Perennial Salt Marsh Aster

Aster tenuifolius L.

Composite or Aster Family
Compositae (Asteraceae)

Description: Medium-height, erect, fleshy perennial herb, from 6 inches to 4 feet tall, with fibrous roots and creeping rhizomes; stems smooth; simple, entire, fleshy, linear, sometimes narrowly lance-shaped leaves (1½–6 inches long), few in number, upper leaves reduced in size, alternately arranged; pale purple or blue or white daisylike flowers in heads (½–1 inch wide) with fifteen to twenty-five petallike rays and a yellow or reddish central disk, several to many heads in an open inflorescence, sometimes solitary.

Flowering period: June through November.

Habitat: Irregularly flooded salt and brackish marshes, tidal flats, and mangrove swamps; marl prairies and limerock pinelands.

Wetland indicator status: OBL.

Range: New Hampshire to Florida and Texas.

Similar species: Lowland White Aster (*A. simplex*) has white daisylike flowers with twenty to forty petallike rays and is found in tidal fresh marshes along the Atlantic Coast, not in salt or brackish marshes; it is FACW. Frost Aster (*A. pilosus*) is similar to *A. simplex* in form and habitat preference, but it has sharp-pointed bracts beneath petallike rays; it is FAC−.

Annual Salt Marsh Aster

Perennial Salt Marsh Aster

Semaphore Thoroughwort

Eupatorium mikanioides Chapm.

Composite or Aster Family
Compositae (Asteraceae)

Description: Medium-height, perennial, somewhat fleshy-leaved herb, erect or reclining, up to 3½ feet tall or long; stems smooth below, hairy above; simple, round-toothed, somewhat fleshy, somewhat triangle-shaped leaves (up to 2¾ inches long and to 2 inches wide) on hairy stalks (up to 1 inch long), oppositely arranged; numerous white flowers in heads borne in a terminal inflorescence.

Flowering period: Summer.

Habitat: Salt, brackish, and tidal fresh marshes; interdunal swales, nontidal marshes, pine flatwoods, limestone sinkholes, ditches, and spoil banks.

Wetland indicator status: FACW.

Range: Daytona Beach, Florida, south to the Keys and north along Florida's Gulf coast to Panama City.

Similar species: Late-flowering Thoroughwort or Boneset (*E. serotinum*) has a flat-topped inflorescence of similar whitish flowers, and its leaves are coarse-toothed and lance-shaped and not fleshy; it is found in brackish and tidal fresh marshes; it is FAC. Mistflower (*Conoclinium coelestinum*, formerly *Eupatorium coelestinum*) has nonfleshy leaves and violet or light purple flowers and occurs in tidal fresh marshes; it is FAC.

Seaside Goldenrod

Solidago sempervirens L. var. *mexicana* (L.) Fernald

Composite or Aster Family
Compositae (Asteraceae)

Description: Medium-height to tall, erect, fleshy perennial herb, usually 4–5 feet high but up to 7 feet; stems smooth but may be rough-hairy in inflorescence; simple, entire, thick, fleshy sessile leaves (4–16 inches long), lance-shaped or oblong, decreasing in size toward top of stem, alternately arranged; numerous yellow flowers in heads with seven to seventeen rays (less than ⅕ inch long) borne on terminal inflorescences with many ascending branches.

Flowering period: August through October.

Habitat: Irregularly flooded salt, brackish, and tidal fresh marshes; sand dunes and beaches.

Wetland indicator status: FACW.

Range: Southeastern Massachusetts to Florida and Texas.

Similar species: Willow-leaf Goldenrod (*S. stricta*) is quite similar but has long, slender rhizomes, unbranched stems; its leaves are not fleshy and they are markedly reduced in size and closely appressed to the stem as they approach the inflorescence; it is OBL. Elliott's Goldenrod (*S. elliottii*) also occurs in brackish marshes as well as freshwater swamps from Nova Scotia to Florida; its leaves are toothed and not fleshy; it is OBL. Other salt marsh goldenrods (*Euthamia graminifolia* and *E. galetorum*) have grasslike linear leaves that are not fleshy.

Semaphore Thoroughwort

lower
leaf

Seaside Goldenrod

Sea Rocket

Cakile edentula (Bigel.) Hook.

Mustard Family
Cruciferae (*Brassicaceae*)

Description: Low-growing, erect, fleshy annual herb, up to 12 inches tall; stems much branched, sometimes creeping; simple, weakly lobed or toothed, sometimes almost entire, fleshy leaves (up to 2 inches long) often somewhat spoon-shaped, narrowing at the base, alternately arranged; pale purple to white four-petaled flowers (¼ inch wide); fruit two-jointed pod with one or two seeds. (*Note:* Fleshy leaves have a mild horseradish taste.)

Flowering period: April through October.

Habitat: Upper zone of coastal beaches and upper elevations of salt marshes (usually associated with tidal wrack).

Wetland indicator status: FACU.

Range: Labrador to South Carolina; inland at the head of Lake Michigan; also reported in Florida.

Similar species: Other sea rockets (*C. constricta*, *C. geniculata*, and *C. lanceolata*) occur on coastal sands in Florida and along the Gulf Coast.

Southern Seaside Arrow Grass

Triglochin striata Ruiz & Pavon

Arrow Grass Family
Juncaginaceae

Description: Fleshy-leaved perennial herb, up to 10 inches tall, with underground stolons; linear, fleshy basal leaves (to 8 inches long and ⅗ inch wide), sheathing at base and roundish in cross-section; inconspicuous greenish flowers borne on narrow, cylinder-shaped, spikelike terminal inflorescence (raceme, up to 10 inches tall and usually shorter than leaves); three-parted, somewhat roundish, triangular fruit capsules.

Flowering period: May through September.

Habitat: Salt and brackish marshes.

Wetland indicator status: OBL.

Range: Maryland and Delaware south to Florida, west to Louisiana; also on West Coast (California and Oregon) and in tropical America.

Similar species: Northern Seaside Arrow Grass (*T. maritimum*) occurs in similar habitats from Maryland north; its flowering inflorescence is usually taller than its leaves; it is OBL. (*Note:* This is the illustrated species.)

Sea Beach Knotweed

Polygonum glaucum Nutt.

Buckwheat or Smartweed Family
Polygonaceae

Description: Low-growing, trailing, fleshy annual or perennial herb, up to 12 inches long; stems jointed and much branched; simple, entire, fleshy, narrowly lance-shaped bluish green leaves (up to 1 inch long and up to ¼ inch wide), margins often rolled backward, upper leaves reduced in size, stalked, alternately arranged, leaf sheaths (ocrea) silvery to whitish above and brownish below; small white to pink flowers (about ¹⁄₁₀ inch long) borne singly or in twos or threes from upper leaf axils; smooth, shiny reddish brown, three-sided nutlets.

Flowering period: May into October.

Habitat: Sandy upper edges of salt marshes; coastal beaches.

Wetland indicator status: FACU.

Range: Massachusetts south to Florida.

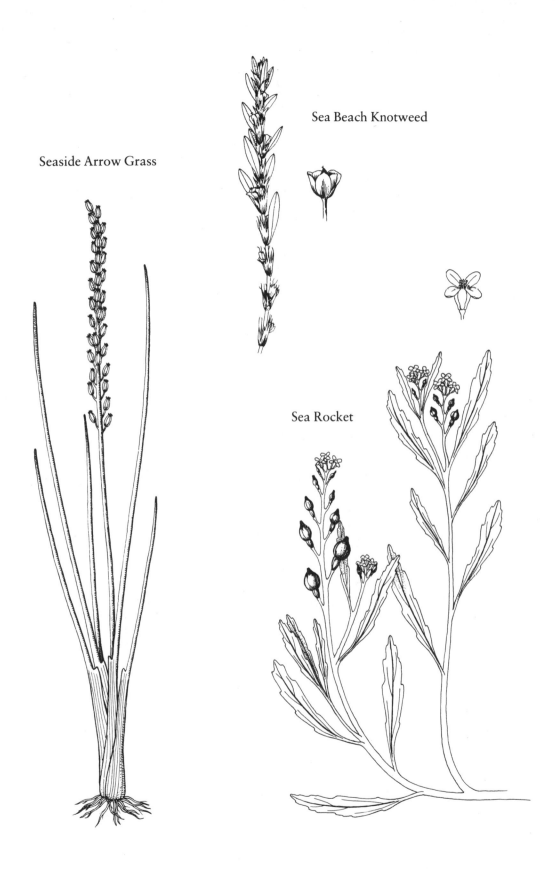

Seaside Arrow Grass

Sea Beach Knotweed

Sea Rocket

Seaside Gerardia

Agalinis maritima Raf.
[*Gerardia maritima* Raf.]

Figwort Family
Scrophulariaceae

Description: Low and medium-height, erect, annual fleshy herb, often 4 inches tall but sometimes to 24 inches; fleshy leaves simple, entire, and linear (up to 1¼ inches long and up to ⅛ inch wide), mostly oppositely arranged but may be alternately arranged on end of branches; small pink to purple five-lobed tubular flowers (½ inch diameter, ⅜–¾ inch long) borne in two to five pairs on stalks (pedicels); calyx lobes blunt with rectangular spaces between. (*Note:* Size, number of branches, and number and size of flowers increase north to south.)

Flowering period: May to October.

Habitat: Irregularly flooded salt and brackish marshes (often in pannes); interdunal swales.

Wetland indicator status: OBL.

Range: Nova Scotia to Florida and Texas.

Similar species: Purple Gerardia (*A. purpurea*, formerly *Gerardia purpurea*) occurs in interdunal swales and other open nontidal wetlands; it is a much-branched annual up to 4 feet tall, with hairy or slightly rough angled stems and many rose-purple flowers (¾–1½ inches long) borne on short stalks; it is FACW.

Coastal Water-hyssop

Bacopa monnieri (L.) Wettst.

Figwort Family
Scrophulariaceae

Description: Fleshy, creeping to erect perennial herb, up to 1 foot tall, often forming dense mats; smooth fleshy stems; simple, entire, fleshy leaves (to ⅗ inch long and ⅓ inch wide), broadest above middle, rounded tips, usually one-veined (rarely two- or three-veined), stalkless, oppositely arranged, internodes between pairs of leaves longer than leaves themselves; small, white, five-"petaled" tubular flowers (about ¼ inch wide and sometimes blue-tinged) borne singly on stalks (up to 1 inch long) in the axil of one of the pair of leaves; oval to egg-shaped fruit capsule.

Flowering period: April to November.

Habitat: Sandy brackish and tidal fresh marshes, sand flats, shallow water; interdunal swales, nontidal marshes, and sandy margins of streams and ponds.

Wetland indicator status: OBL.

Range: Southeastern Virginia along the Coastal Plain to the Florida Keys, west to Texas; also in subtropical and tropical America.

Similar species: Carolina Water-hyssop (*B. caroliniana*) is aromatic (lemon-scented when crushed) with hairy stems and has three- to seven-veined leaves that are broadest below the middle and clasping the stem at their bases. Disk Water-hyssop (*B. rotundifolia*) has roundish leaves with clasping bases; its stems are hairy, and its five-"petaled" flowers are whitish and yellowish. Shade Mudflower (*Micranthemum umbrosum*) also has roundish leaves (from ⅕ to ½ inch wide); its whitish flowers are minute (less than ¹⁄₁₀ inch long) and four-"petaled." All these species are OBL.

SEE ALSO Annual Salt Marsh Fleabane (*Pluchea purpurascens*), Narrow-leaved Cattail (*Typha angustifolia*), and fleshy-leaved shrubs.

Coastal Water-hyssop

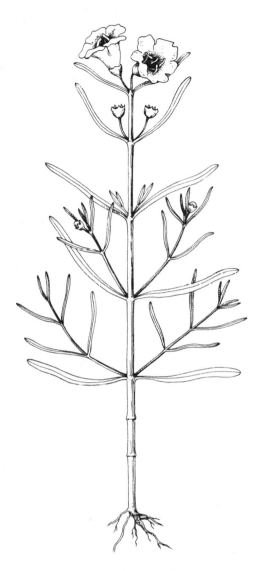

Seaside Gerardia

FLOWERING HERBS WITH BASAL LEAVES ONLY

Northern Sea Lavender or Marsh Rosemary

Limonium nashii Small

Leadwort Family
Plumbaginaceae

Description: Low, erect, perennial herb with flowering inflorescence, up to 3 feet tall; simple, entire, basal leaves (2–6 inches long), lance-shaped to spoon-shaped, tapering at base into an often red-tinged petiole usually longer than leaves; numerous minute, bluish to lavender, five-lobed tubular flowers borne on a single, tall inflorescence arising from basal leaves and widely branched above the middle.

Flowering period: June through November.

Habitat: Irregularly flooded salt marshes, mangrove swamps, and salt flats; interdunal swales.

Wetland indicator status: OBL.

Range: Labrador and Quebec south to Florida and northeastern Mexico.

Similar species: Carolina Sea Lavender (*L. carolinianum*) occurs from southern New York south; it has smooth flowers, whereas *L. nashii* has fine-hairy flowers (hairy at least at base); it is OBL.

Marsh Pennywort

Hydrocotyle umbellata L.

Parsley Family
Umbelliferae

Description: Low-growing perennial herb, usually to 6 inches tall but ranging to 10 inches; stems rooting at nodes; simple, thick, round-toothed, roundish basal leaves (up to 1⅗ inches wide), long stalks (up to 10 inches) attached to center of leaf (peltate); many (fifteen to thirty) inconspicuous whitish or greenish flowers borne on separate, long-stalked (equaling or exceeding leaf stalk), branched terminal inflorescence (simple umbel).

Flowering period: April through September.

Habitat: Sandy upper edges of salt and brackish marshes and tidal fresh marshes; pond and lake shores and nontidal marshes.

Wetland indicator status: OBL.

Range: Nova Scotia to Florida, west to Minnesota and Texas; also on the West Coast and in tropical America.

Similar species: Coastal Plain Pennywort (*H. bonariensis*) has flowers arranged in whorls along a many-branched umbel; it is FACW. Whorled Pennywort (*H. verticillata*) has flower stalks that are shorter than the leaf stalks and has two to seven flowers borne in whorled clusters; it is OBL. Asiatic Coinleaf (*Centella asiatica*, formerly *C. erecta*) lacks peltate leaves, its leaf and flower stalks are usually hairy, its leaf stalks are sheathed, and its leaves are oblong to somewhat heart-shaped; it is FACW.

Northern Sea Lavender flower

Carolina Sea Lavender flower

Northern Sea Lavender

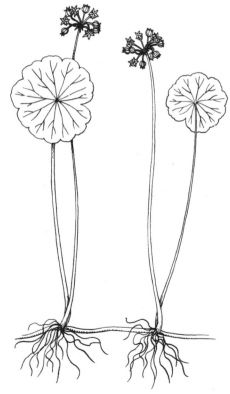

Marsh Pennywort

Eastern Lilaeopsis

Lilaeopsis chinensis (L.) Kuntze

Parsley Family
Umbelliferae

Description: Very low, erect perennial herb, up to 2½ inches tall, growing from creeping rhizome; stems rhizomatous; simple, flattened, linear basal "leaves" (up to 2 inches long) with four to six transverse septa ("leaves" are actually phyllodes—a flattened petiole without a leaf blade); very small white flowers borne in an umbel on a separate stalk longer than the "leaves."

Flowering period: April to September.

Habitat: Brackish and tidal fresh marshes (regularly and irregularly flooded zones) and tidal mud flats.

Wetland indicator status: OBL.

Range: Nova Scotia to Florida and Mississippi.

Similar species: Carolina Lilaeopsis (*L. attenuata*, formerly *L. carolinensis*) has longer leaves (to 12 inches) that exceed shorter flowering stalks (2 inches tall or less); it is FACW.

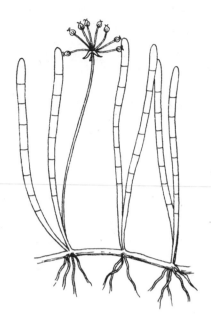

Eastern Lilaeopsis

SEE ALSO Mudwort (*Limosella subulata*).

FLOWERING HERBS WITH SIMPLE, ENTIRE, ALTERNATE LEAVES

Water Hemp

Amaranthus cannabinus (L.) Sauer
[*Acnida cannabina* L.]

Amaranth Family
Amaranthaceae

Description: Tall, erect annual herb, up to 8 feet high; stems smooth; simple, entire, lance-shaped or linear (uppermost) leaves (up to 6 inches long) on long stalks (petioles, ⅕–2 inches long), alternately arranged; small green or yellow-green flowers borne on slender spikes from leaf axils and terminally.

Flowering period: July to November.

Habitat: Salt, brackish, and tidal fresh marshes (usually regularly flooded zone).

Wetland indicator status: OBL.

Range: Southern Maine to Florida.

Similar species: Southern Water Hemp (*A. australis*, formerly *Acnida alabamensis* or *A. cuspidata*) occurs in similar habitats along the Gulf coast from Florida to Texas and Mexico; it is OBL.

Water Hemp

Grass-leaved Goldenrod

Euthamia graminifolia (L.) Nutt.
[*Solidago graminifolia* (L.) Salisb.]

Composite or Aster Family
Compositae (Asteraceae)

Description: Medium-height, erect perennial herb, 1–4 feet tall; stems branched at top forming a flattish inflorescence; simple, entire, linear leaves (less than ¼ inch wide), sometimes narrowly lance-shaped, three-nerved (larger leaves with four to five veins), alternately arranged; twenty to thirty-five yellow flowers in heads with fifteen to twenty-five small rays, borne in terminal flat-topped inflorescence.

Flowering period: July through October.

Habitat: Brackish and tidal fresh marshes and upper edges of salt marshes; nontidal marshes and meadows and various open moist or dry inland habitats.

Wetland indicator status: FACW−.

Range: Newfoundland and Quebec to Alberta and South Dakota (in mountains to British Columbia), south to Virginia and North Carolina (in mountains to New Mexico).

Similar species: Slender-leaved Goldenrod (*E. galetorum,* formerly *Solidago tenuifolia,* also *Euthamia tenuifolia*) is more common in southern brackish and freshwater marshes and along salt marsh borders; it has one- to three-nerved (usually one-nerved) leaves and five to seven, rarely nine, yellow flowering heads with eight to fifteen ray flowers and five to seven disk flowers, and it often has clusters of smaller leaves in leaf axils; it is FAC.

Grass-leaved Goldenrod

SEE ALSO Seaside Heliotrope (*Heliotropium curassavicum*), Perennial Salt Marsh Aster (*Aster tenuifolius*), Annual Salt Marsh Aster (*A. subulatus*), Water Pimpernel (*Samolus parviflorus*), and Seaside Goldenrod (*Solidago sempervirens*).

FLOWERING HERBS WITH SIMPLE, ENTIRE, OPPOSITE LEAVES

Bloodleaf or Rootstock Bloodleaf
Iresine rhizomatosa Standl.

Amaranth Family
Amaranthaceae

Description: Erect, medium-height perennial herb, up to 5 feet tall; slender, horizontal rhizomes; stems with somewhat swollen nodes; simple, entire, somewhat egg-shaped to lance-shaped leaves (up to 6 inches long) with pointed tips, stalked (to 2½ inches long) or short-winged stalks (on upper leaves), fine-hairy on both sides, oppositely arranged; many minute white flowers borne on narrow, branching spikes (each less than ¾ inch long) forming terminal and axillary inflorescences (panicles, up to 12 inches long); roundish, bladderlike fruit bearing one brownish red seed.

Flowering period: August through October.

Habitat: Upper margins of salt marshes; interdunal swales, nontidal forested wetlands, and low woods.

Wetland indicator status: FACW−.

Range: Maryland south to Florida, west to eastern Texas and Kansas.

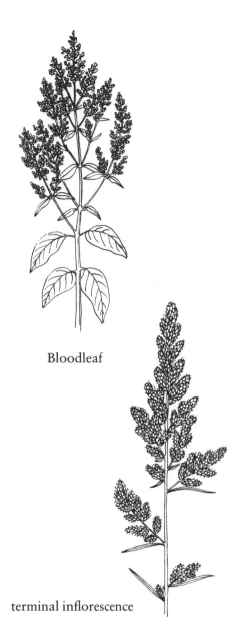

Bloodleaf

terminal inflorescence

Red Milkweed

Asclepias lanceolata Walter

Milkweed Family
Asclepiadaceae

Description: Medium-height perennial herb, up to 4 feet tall; smooth, purplish (especially lower part) stems with tuberous roots; milky sap; simple, entire, linear leaves (up to 8 inches long and to ⅗ inch wide) tapering to a point, oppositely arranged in three to six pairs; several red, orange, or reddish purple regular flowers (about ½ inch wide) composed of five erect hoods and five downward-pointing lobes borne on stalks in one to four terminal inflorescences (umbels, up to 2 inches wide); smooth, elongate fruit pod (follicle, up to 4 inches long and less than ½ inch wide) borne on stalks and bending downward.

Flowering period: May through August.

Habitat: Brackish and tidal fresh marshes; nontidal marshes, wet pine barrens, savannahs, and forested wetlands along the Coastal Plain.

Wetland indicator status: OBL.

Range: Southern New Jersey to Florida, west to southeastern Texas.

Similar species: Aquatic Milkweed (*A. perennis*) has white or pink flowers and lance-shaped leaves; it is OBL. Swamp Milkweed (*A. incarnata*) occurs in tidal fresh marshes and along the edges of brackish marshes (see description); it is OBL. Scarlet Milkweed (*A. currassavica*) is an annual with somewhat woody stem bases, broader leaves (to 1¼ inches wide), and usually bright red flowers (rarely yellow or white); it is FAC.

Perennial or Large Salt Marsh Pink

Sabatia dodecandra (L.) B.S.P.

Gentian Family
Gentianaceae

Description: Low to medium-height, erect perennial herb, 1–2½ feet tall, lacking stolons; stems bearing alternate branches above middle; simple, entire, sessile, lance-shaped leaves (⅘–2 inches long), oppositely arranged; pink, sometimes white, regular flowers with eight to twelve petals and yellow center outlined in red (1½–2½ inches wide), terminal flower typically long-stalked.

Flowering period: June into September.

Habitat: Irregularly flooded salt and brackish marshes, rarely tidal fresh marshes; pond and river margins, ditches, nontidal marshes, and pine savannahs.

Wetland indicator status: OBL.

Range: Connecticut and Long Island south to Florida, Louisiana, and Texas.

Similar species: Plymouth Rose-gentian (*S. kennedyana*) is a rare species with stolons, flowers with nine to twelve petals, usually short-stalked and with some of its primary branches oppositely arranged; it occurs in North and South Carolina and sporadically farther north; it is OBL. The other common marsh pinks (*S. stellaris* and *S. campanulata*) have only five or six petals. Catchfly Gentian (*Eustoma exaltatum*) is a Gulf coast annual with blue, lavender, or white five- or six-petaled flowers lacking yellow centers; it is FACW+.

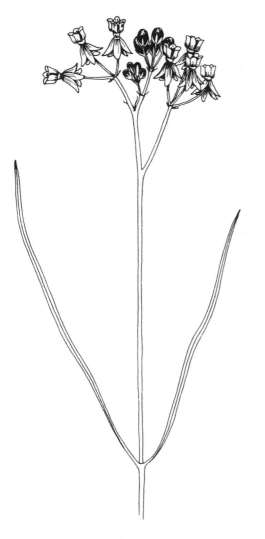

Red Milkweed

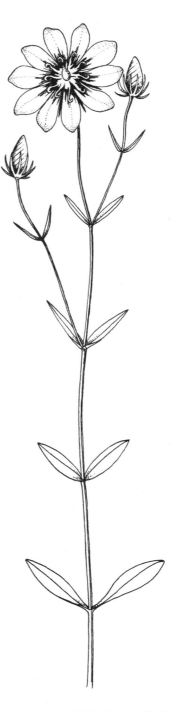

Perennial Salt Marsh Pink

Annual or Small Salt Marsh Pink

Sabatia stellaris Pursh

Gentian Family
Gentianaceae

Description: Low to medium-height, erect annual herb, 4–24 inches tall; simple, entire, sessile leaves, linear to egg-shaped, narrow at base (1–1½ inches long), oppositely arranged; pink five-petaled regular flowers with yellow center (¾–1½ inches wide), calyx lobes shorter than petals.

Flowering period: July into October.

Habitat: Irregularly flooded salt and brackish marshes, especially sandy borders of salt marshes; interdunal swales.

Wetland indicator status: OBL.

Range: Southeastern Massachusetts to Florida, west to Louisiana; also in the Bahamas, Cuba, and Mexico.

Similar species: Slender Marsh Pink (*S. campanulata*) occurs in salt and brackish marshes in the same range; it is FACW; its leaves are rounded at base, and calyx lobes are as long as petals, whereas *S. stellaris* has leaves narrowing at base and calyx lobes shorter than petals. Perennial Salt Marsh Pink (*S. dodecandra*) has eight to twelve petals; it is OBL.

Salt Marsh Loosestrife

Lythrum lineare L.

Loosestrife Family
Lythraceae

Description: Medium-height, erect perennial herb, up to 5 feet tall, commonly about 3–4 feet; four-angled (squarish) stems; simple, entire, somewhat linear to lance-shaped leaves (up to 1⅗ inches long and ⅕ inch wide; smaller leaves above) with pointed tips, wedged-shaped bases, stalkless, oppositely arranged; small, four- to six-"petaled," white or light purplish tubular flowers (about ¼ inch wide) borne singly in leaf axils of flowering branches; somewhat cylinder-shaped fruit capsules (less than ⅕ inch long).

Flowering period: July through October.

Habitat: Irregularly flooded salt, brackish, and tidal fresh marshes.

Wetland indicator status: OBL.

Range: Long Island, New York, south along the Coastal Plain to Florida, west to Texas.

SEE ALSO Seaside Gerardia (*Agalinis maritima*).

Salt Marsh Loosestrife

Slender Marsh Pink flower

Annual Salt Marsh Pink

FLOWERING HERBS WITH SIMPLE, TOOTHED, ALTERNATE LEAVES

Annual Salt Marsh Fleabane

Pluchea purpurascens (Swartz) DC.
[includes *Pluchea odorata* (L.) Cass.]

Composite or Aster Family
Compositae (Asteraceae)

Description: Low to medium-height, erect annual herb, usually 8–36 inches tall but up to 5 feet high; simple, sharply toothed (sometimes obscurely toothed to entire), hairy, aromatic leaves (1½–6 inches long), egg-shaped or lance-shaped with petioles or tapering to the base, alternately arranged; pink or purple flowers in heads (¼ inch long) borne in flat-topped or somewhat rounded inflorescences, bracts hairy.

Flowering period: August through October.

Habitat: Irregularly flooded salt and brackish marshes, occasionally tidal fresh marshes (sometimes regularly flooded zone); interdunal wet swales and nontidal marshes.

Wetland indicator status: FACW+.

Range: Southern Maine to Florida, west to California; occasionally inland, as in western New York, Michigan, and Kansas.

Similar species: Marsh Fleabane (*P. foetida*) occurs from southern New Jersey south in freshwater wetlands near the coast; it is perennial with sessile leaves that clasp the stem and creamy white flowers; it is FACW. Long-leaf Camphorweed (*P. longifolia*) is very similar to *P. foetida* and occurs in brackish and tidal fresh marshes in northern and central Florida; its flower heads are longer—twice as long as wide; it is OBL. Camphorweed (*P. camphorata*) occurs from Delaware south; it is quite similar to *P. purpurascens*, but its leaves and stems are usually darker green and smooth, and its bracts are smooth or glandular; it is OBL.

Rose or Marsh Mallow

Hibiscus moscheutos L.
[*Hibiscus palustris* L. and *Hibiscus lasiocarpos* Cav.]

Mallow Family
Malvaceae

Description: Tall, erect perennial herb, 3½–7 feet high; stems round, hairy above and smooth below; simple, toothed leaves egg-shaped or sometimes obscurely three-lobed with rounded or heart-shaped bases, smooth above, fine-hairy below, alternately arranged; large, showy pink or white five-petaled flowers (4–6½ inches wide) with purple or red centers; five-celled fruit capsule, rounded at top.

Flowering period: May through September.

Habitat: Irregularly flooded salt, brackish, and tidal fresh marshes; nontidal marshes.

Wetland indicator status: OBL.

Range: Massachusetts to Florida and Alabama; inland from western New York and southern Ontario to northern Illinois and Indiana.

Similar species: Swamp Rose Mallow (*H. grandiflorus*) has larger flowers with pink petals (about 5–5½ inches long) and hairy fruit capsules, whereas *H. moscheutos* has petals about 4 inches long; Swamp Rose Mallow is OBL. Halberd-leaved Rose Mallow (*H. laevis*, formerly *H. militaris*) usually has strongly three-lobed leaves that are smooth on both surfaces; it is OBL. Scarlet Rose Mallow (*H. coccineus*) is smooth (not hairy) throughout stems and leaves and its flowers are deep red; it occurs from southeastern Georgia to Mississippi; it is OBL. Mangrove Mallow (*Pavonia spicata*) has smaller, greenish yellow flowers (less than 2 inches wide); it occurs in mangrove swamps in southern Florida; it is OBL.

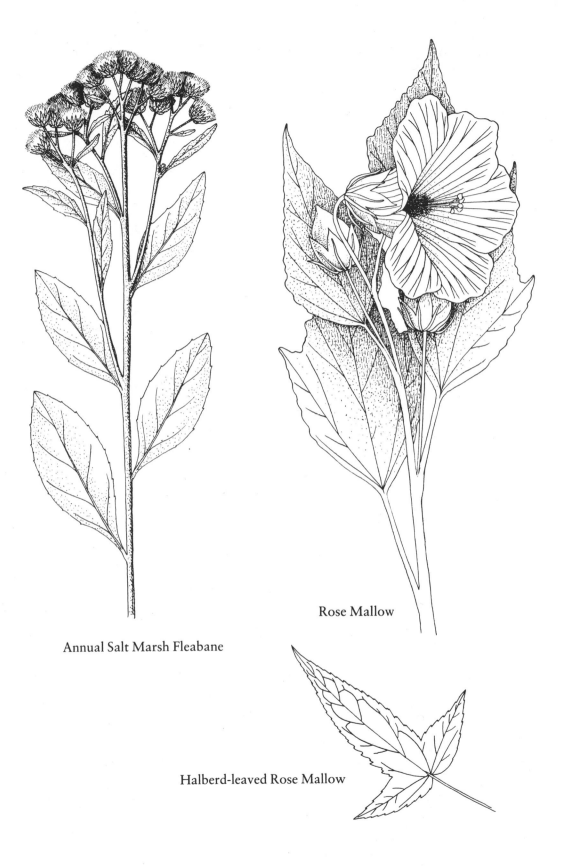

Annual Salt Marsh Fleabane

Rose Mallow

Halberd-leaved Rose Mallow

Seashore Mallow

Kosteletzkya virginica (L.) K. Presl ex Gray

Mallow Family
Malvaceae

Description: Medium-height, erect perennial herb, up to 4 feet tall; stems round and rough-hairy; simple, coarsely toothed, rough-hairy leaves (2½–6 inches long) generally triangular egg-shaped, usually with three pointed lobes, alternately arranged; pink (rarely white) five-petaled flowers (1½–2½ inches wide) in leaf axils and a terminal spike; five-celled, flattened globe-shaped fruit capsule.

Flowering period: June through October.

Habitat: Irregularly flooded salt, brackish, and tidal fresh marshes; nontidal marshes.

Wetland indicator status: OBL.

Range: Long Island along the Coastal Plain to Florida and Texas.

Similar species: White Fen-rose (*K. depressa*, formerly *K. pentasperma*) has white flowers and occurs in southern Florida on the edges of mangrove swamps or in coastal hammocks; it is FAC+.

Seashore Mallow

SEE ALSO New York Aster (*Aster novi-belgii*) and Elongated Lobelia (*Lobelia elongata*).

FLOWERING HERBS WITH SIMPLE, TOOTHED, OPPOSITE LEAVES

Late-flowering Thoroughwort or Boneset

Eupatorium serotinum Michx.

Composite or Aster Family
Compositae (Asteraceae)

Description: Medium-height to tall, erect perennial herb, up to 6½ feet high, often forming clumps; stem hairy, mostly solid; simple, round- or sharp-toothed, lance-shaped to egg-shaped leaves (up to 5 inches long and to 2½ inches wide) with two prominent lateral veins parallel to midrib, leaf undersides hairy with resin dots, stalked, oppositely arranged, upper leaves sometimes alternately arranged; small white to light purplish flowers (less than ⅕ inch long) borne in ten to fifteen clusters (corymbs, less than 1¾ inches wide) forming a somewhat flat-topped inflorescence; nutlets black, somewhat sticky.

Flowering period: August through October.

Habitat: Brackish marshes (especially Black Needlerush-dominated) and tidal fresh marshes; nontidal marshes, floodplain forests, stream banks, upland fields, and waste places.

Wetland indicator status: FAC.

Range: Massachusetts to Wisconsin, south to Florida and Texas; also in Mexico.

Similar species: May hybridize with Boneset (*E. perfoliatum*).

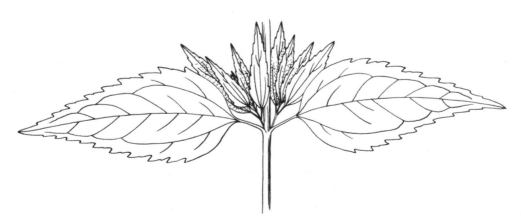

Late-flowering Thoroughwort lower leaves

American Germander

Teucrium canadense L.

Mint Family
Labiatae (Laminaceae)

Description: Medium-height, erect perennial herb, 1–4 feet tall; stems square and hairy (sometimes smooth), rarely branched; sharply or obscurely toothed, simple leaves, oblong to lance-shaped (2–5 inches long), on short stalks (petioles, ¹⁄₁₆–½ inch long), oppositely arranged; small pink-purple or creamy white tubular flowers (⅜–1⅛ inches long) with a broad lower lip (upper lip absent) borne in dense terminal spike.

Flowering period: June through August.

Habitat: Upper edges of salt marshes, irregularly flooded brackish and tidal fresh marshes; inland shores, woods, thickets, and moist or wet soils.

Wetland indicator status: FACW−.

Range: New Brunswick and Nova Scotia south to Florida and Texas, west to Minnesota and Oklahoma.

Common Frog-fruit

Phyla nodiflora (L.) Greene
[*Lippia nodiflora* (L.) Michx.]

Vervain Family
Verbenaceae

Description: Short perennial herb, usually less than 4 inches tall, with creeping stems that root at nodes; simple, toothed (usually three to five pairs of teeth, rarely seven, mostly above middle of leaf), somewhat egg-shaped to spoon-shaped leaves (up to 1⅓ inches long and to ⅘ inch wide) with blunt tips and wedge-shaped bases, oppositely arranged; small pinkish, light purplish, or whitish, five-lobed tubular flowers (less than ⅕ inch wide) borne in a dense terminal headlike spike (about 1 inch long) at end of long stalk (up to 4 inches long); fruit capsule divided into two nutlets.

Flowering period: Late May to November.

Habitat: Sandy edges of irregularly flooded salt marshes, and tidal fresh marshes; open sandy areas in dunes.

Wetland indicator status: FACW.

Range: Southeastern Virginia along the Coastal Plain to Florida, west to Texas; inland from southwestern Missouri and Arkansas to Oklahoma; also in southern California and tropical America.

Similar species: Lance-leaved Frog-fruit (*P. lanceolata*, formerly *Lippia lanceolata*) is taller (8–32 inches); its leaves are much longer than wide (to 2½ inches long) and have seven to eleven (rarely five) pairs of teeth along the margins; it is FACW+.

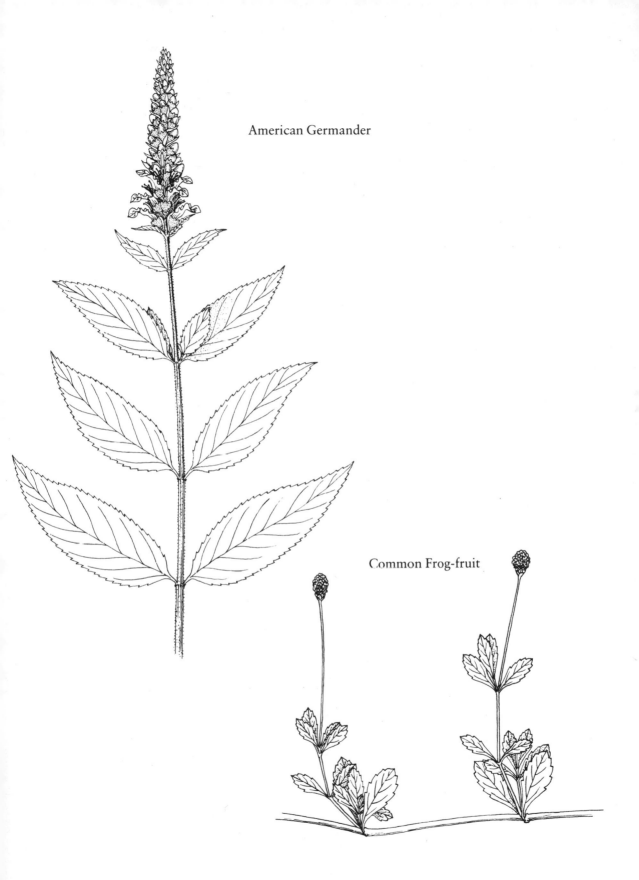

American Germander

Common Frog-fruit

Sandpaper Vervain

Verbena scabra Vahl

Vervain Family
Verbenaceae

Description: Medium-height, erect perennial herb, up to 5 feet tall; stiff-hairy, squarish stems with many branches; simple, round- to coarse-toothed leaves (up to 4¾ inches long and to 2½ inches wide) with pointed tips and rounded to wedge-shaped bases, very rough upper leaf surfaces, less rough lower surfaces, stalked, oppositely arranged; small pinkish to lavender, five-lobed tubular flowers (about ⅛ inch wide) borne in open terminal spikes (up to 8 inches long); small oblong nutlets.

Flowering period: March to December.

Habitat: Margins of brackish and tidal fresh marshes and on shell deposits in coastal marshes; nontidal marshes, swamps, and edges of lakes and streams.

Wetland indicator status: FACW+.

Range: Southeastern Virginia to Florida, west to California; also in Mexico and the West Indies.

Similar species: Blue Vervain (*V. hastata*) may occur in tidal fresh marshes; its bluish to violet (rarely pinkish) flowers are borne in dense spikes; it is FAC.

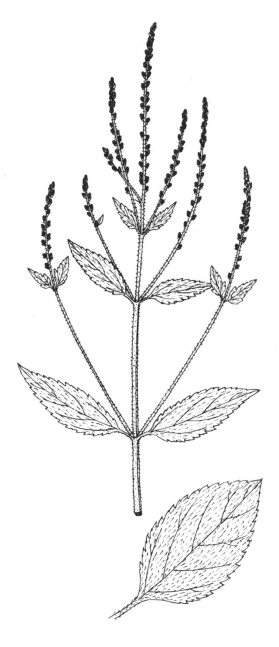

Sandpaper Vervain

Blue Vervain

FLESHY-LEAVED SHRUBS

Saltwort

Batis maritima L.

Saltwort Family
Bataceae

Description: Fleshy-leaved woody shrub, up to
3¼ feet tall, often less than 1 foot tall, with trailing, arching, and erect stems rooting at nodes;
young stems succulent, older stems woody with
flaky bark; simple, entire, linear to club-shaped,
yellowish green, fleshy leaves (up to 1¼ inches
long, half-round in cross-section), oppositely arranged; male and female flowers borne on axillary spikes on separate plants (dioecious), male
flowers with many overlapping scales and stalkless, female flowers short-stalked; fleshy fruits
fused together forming roundish clusters of two
to eight fruits (about ½ inch long).

Flowering period: June through July.

Habitat: Irregularly flooded salt marshes, sand
flats and barrens, and mangrove swamps.

Wetland indicator status: OBL.

Range: South Carolina to the Florida Keys, west
to Texas; also in the Caribbean, tropical America, southern California, and Hawaii.

Similar species: Perennial Glasswort (*Salicornia
virginica*) occurs in similar habitats and has
fleshy, woody stems, but its leaves are reduced
to minute scales; it is OBL.

Sea Ox-eye

Borrichia frutescens (L.) DC.

Composite or Aster Family
Compositae (Asteraceae)

Description: Low-growing, fleshy-leaved deciduous shrub, up to 2½ feet tall, typically forming
dense stands; stem grayish hairy; simple, entire
(rarely spiny-toothed), grayish hairy, fleshy
leaves (up to 3¼ inches long and to 1¼ inches
wide) with sharp-pointed tips and tapering
bases, stalkless, oppositely arranged; medium-
sized, showy, many-"petaled," daisylike yellow
flowers (yellow rays to ⅗ inch long, often with
lobed tips) borne terminally; sharply four-sided,
black nutlets in prickly fruiting heads.

Flowering period: May through September.

Habitat: Irregularly flooded salt marshes, salt
flats, brackish marshes, estuarine shores, and
edges of mangrove swamps.

Wetland indicator status: OBL.

Range: Eastern Virginia and possibly Maryland
(Assateague Island) to the Florida Keys, west
into Texas and northern Mexico; also in Bermuda and the Caribbean.

Similar species: Bay Marigold (*B. arborescens*)
occurs in mangrove swamps and along estuarine
shores of south Florida; it usually does not form
dense stands; it grows up to 4½ feet tall, and its
stems are somewhat prickly, though its fruiting
heads are not; it is FACW.

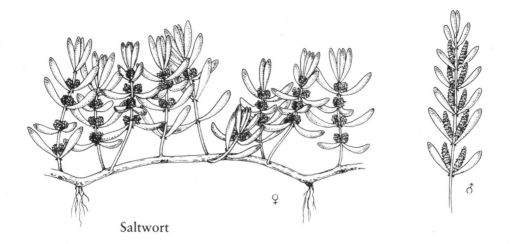

Saltwort

♀

♂

Sea Ox-eye

High-tide Bush or Marsh Elder

Iva frutescens L.

Composite or Aster Family
Compositae (Asteraceae)

Description: Deciduous shrub, 2–12 feet tall, usually less than 6 feet high; stems hairy above and often smooth below; twigs branched with vertical lines; simple, coarse-toothed, somewhat fleshy leaves (up to 4¾ inches long), egg-shaped to narrowly lance-shaped, tapering to a petiole, hairy on both surfaces, oppositely arranged except for uppermost reduced leaves; small greenish white flowers in heads borne on erect leafy spikes.

Flowering period: June through November.

Habitat: Irregularly flooded salt and brackish marshes, especially on mounds next to ditches and along upper borders, and mangrove swamps.

Wetland indicator status: FACW+.

Range: Nova Scotia and southern New Hampshire south to Florida and Texas.

Similar species: Sometimes confused with Groundsel Bush (*Baccharis halimifolia*), which also is common in salt marshes, but *Baccharis* has alternately arranged leaves with coarse teeth above the middle; it is FAC. Seashore Elder (*I. imbricata*) occurs on adjacent beaches and low dunes; its leaves and stems are smooth; it is FACW.

Christmas-berry or Carolina Wolfberry

Lycium carolinianum Walter

Nightshade Family
Solanaceae

Description: Thorny, fleshy-leaved deciduous shrub, up to 10 feet tall; simple, entire, linear, fleshy leaves (up to 1 inch long and to ⅕ inch wide), sometimes with club-shaped tips and narrowing toward base, stalkless, alternately arranged, often with short, leafy branches in leaf axils; medium-sized bluish or light purplish (sometimes whitish), four- to seven-"petaled" tubular flowers (about ½–¾ inch wide) usually borne singly from leaf axils; shiny red berries (up to ¾ inch long) somewhat elongate and tomatolike.

Flowering period: September through October.

Habitat: Irregularly flooded salt and brackish marshes, salt barrens and flats, buttonwood hammocks, and shell beaches and mounds.

Wetland indicator status: FACW.

Range: South Carolina to Florida, west to Texas; also in the West Indies.

SEE ALSO Shrubby Sea Blite (*Suaeda conferta*), described under Sea Blite (*S. linearis*).

High-tide Bush

flower

Christmas-berry with fruits

EVERGREEN SHRUBS WITH BROAD LEAVES

Yaupon

Ilex vomitoria Soland. in Ait.

Holly Family
Aquifoliaceae

Description: Broad-leaved evergreen shrub or low tree, up to 28 feet tall; young twigs variably hairy, older twigs smooth and grayish; simple, round-toothed, somewhat egg-shaped to oval, leathery leaves (up to 1¼ inches long) with shiny, dark green upper surfaces, hairy-stalked, alternately arranged; small whitish, four-lobed flowers (less than ⅕ inch long) borne singly or in groups of two or three in leaf axils or at leaf nodes; shiny, round, bright red (rarely yellow) berry bearing four nutlets.

Flowering period: March through May.

Habitat: Upper edges of salt and brackish marshes; interdunal swales, nontidal forested wetlands, coastal dunes, maritime forests, hydric hammocks, well-drained forests, pine flatwoods, and sandhills.

Wetland indicator status: FAC.

Range: Southeastern Virginia to central Florida, west into Texas and southeastern Oklahoma.

Similar species: See Dahoon (*I. cassine*) and Inkberry (*I. glabra*). Possum-haw (*I. decidua*) has thin, deciduous, round-toothed, gland-tipped leaves with wedge-shaped bases; it is FACW−.

Dahoon

Ilex cassine L.

Holly Family
Aquifoliaceae

Description: Broad-leaved evergreen shrub or short tree, up to 36 feet tall; twigs hairy; simple, entire or variously toothed, egg-shaped to lance-shaped evergreen leaves (up to 4 inches long and to 1½ inches wide), stalked (grayish hairy), dark green upper surfaces, pale and usually hairy below, alternately arranged; small, four- to six-"petaled," white flowers (⅕ inch or less wide) borne in clusters in leaf axils or just below the nodes, separate male and female flowers borne on separate plants (dioecious); reddish (sometimes yellowish or orange) round berry (about ¼–⅓ inch wide).

Flowering period: May through June.

Habitat: Edges of brackish marshes; nontidal swamps, bogs, pocosins, hydric hammocks, stream banks, and pond shores.

Wetland indicator status: FACW.

Range: Southeastern Virginia to Florida, west to southeastern Texas.

Similar species: Myrtle Holly (*I. myrtifolia*) has narrow leaves (up to 1¾ inches long and about ¼ inch wide); it is regarded by some authors as a variety of *I. cassine*; it is FACW. Large or Sweet Gallberry (*I. coriacea*) bears black berries and, unlike *I. cassine*, has main veins visible on the undersides of its leaves; a few short, pricklelike marginal teeth are often scattered mostly above the middle of the leaf; it is FACW.

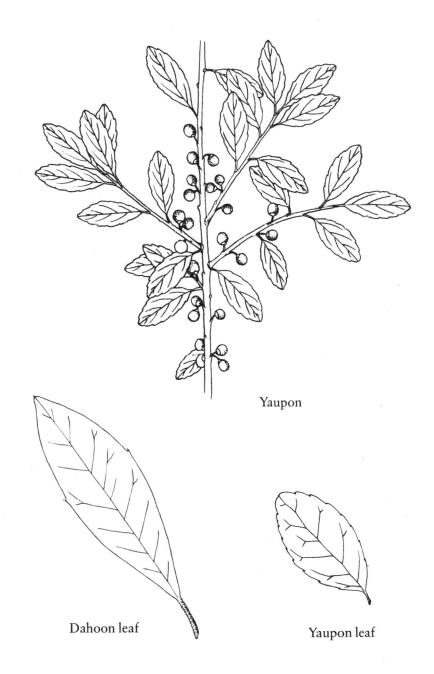

Yaupon

Dahoon leaf

Yaupon leaf

Coin-vine

Dalbergia ecastophyllum (L.) Taubert
Legume Family
Leguminosae

Description: Vinelike, broad-leaved, evergreen tropical shrub with spreading branches reclining over other plants, up to 20 feet tall; older twigs smooth and brown with conspicuous raised lenticels; simple, entire, leathery, somewhat egg-shaped leaves (up to 5 inches long and to about 3 inches wide) with tapered tips and somewhat rounded bases, stalked, alternately arranged; small, white or pink, fragrant, pealike flowers borne in short axillary clusters (panicles); flattened, roundish, coin-shaped fruit (about 1 inch wide).

Flowering period: Yearlong, mostly in spring.

Habitat: Upper edges of mangrove swamps; coastal dune thickets and hammocks.

Wetland indicator status: FACW+.

Range: Central Florida to the Keys; also in the West Indies, tropical America, and west coast of Africa.

Similar species: Brown's Dalbergia (*D. brownei*) occurs in similar habitats; it has compound leaves divided into two to seven leaflets on young plants and its fruits are oblong (longer than wide) and compressed; it is FACW.

Wax Myrtle

Myrica cerifera L.
Bayberry Family
Myricaceae

Description: Medium-height to tall, broad-leaved evergreen shrub or tree, up to about 36 feet high, usually 10–15 feet; bark smooth and grayish green; twigs waxy with few hairs or hairless; simple, entire or weakly coarse-toothed (above middle), oblong to oblong-lance-shaped evergreen leaves (up to 3⅖ inches long and 1 inch wide), yellow-green, shiny, leathery, aromatic when crushed (bayberry scent), base of leaf wedge-shaped, covered with resin dots (glands) above and below, short-petioled, alternately arranged; two types of flowers (male and female) borne in clusters (catkins) in leaf axis, male catkins oval and female catkins linear; round and waxy fruit ball (drupe).

Flowering period: March into June.

Fruiting period: Summer through winter.

Habitat: Irregularly flooded tidal fresh marshes and swamps, occasionally forming dense thickets, upper edges of salt and brackish marshes, and sandy dune swales; nontidal swamps (near coast) and hydric hammocks.

Wetland indicator status: FAC+.

Range: Southern New Jersey south along the Coastal Plain to Florida, west to Texas and Oklahoma.

Similar species: Evergreen Bayberry (*M. heterophylla*) has blackish, soft-hairy twigs that are not very waxy; its leaves are not covered by many resin dots on upper surfaces; it is FACW.

SEE ALSO Red Mangrove (*Rhizophora mangle*), Black Mangrove (*Avicennia germinans*), White Mangrove (*Laguncularia racemosa*), and Buttonwood (*Conocarpus erectus*).

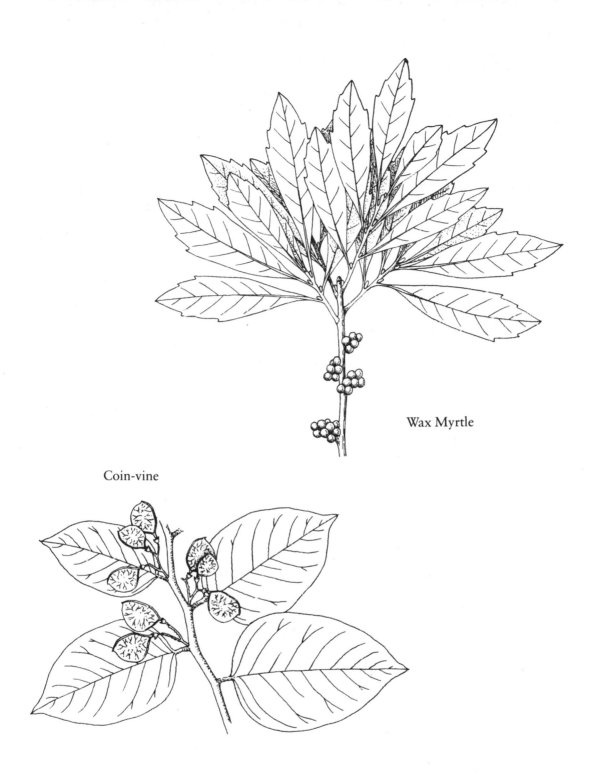

Wax Myrtle

Coin-vine

EVERGREEN SHRUBS WITH NEEDLELIKE OR SCALELIKE LEAVES

Eastern Red Cedar

Juniperus virginiana L.

Pine Family
Pinaceae

Description: Scale-leaved and needle-leaved evergreen coniferous shrub or tree, up to 60 feet tall; reddish brown shaggy bark; four-angled twigs; two types of dark green leaves (to ⅛ inch long)—(1) triangle-shaped leaves flattened, overlapping in four rows covering twigs, and (2) needlelike sharp-pointed leaves, pungent odor when crushed, oppositely arranged; wax-covered whitish green to purplish blue berrylike fruits.

Flowering period: Late January into March.

Fruiting period: Summer and fall.

Habitat: Borders of salt and brackish marshes; nontidal forested wetlands, dry upland woods, and abandoned fields and pastures.

Wetland indicator status: FACU−.

Range: Southern Quebec and Maine to North Dakota, south to Florida and Texas.

Similar species: Southern Red Cedar (*J. silicicola*) is very similar and often misidentified as *J. virginiana*; its berries are smaller (less than ⅕ inch long), whereas those of the latter species are ⅕ inch or longer; Southern Red Cedar is FAC. French Tamarisk (*Tamarix gallica*), an introduced shrub or small tree with scalelike leaves, may occur along borders of salt and brackish marshes; its leaves are grayish green and it produces many very small pinkish to whitish flowers on spikelike inflorescences (racemes) clustered together to form a terminal inflorescence (panicle); it is FACW.

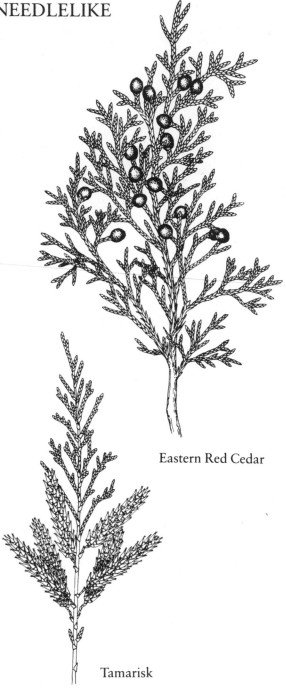

Eastern Red Cedar

Tamarisk

SEE ALSO Australian Pine (*Casuarina equisetifolia*), described under Loblolly Pine (*Pinus taeda*).

DECIDUOUS SHRUBS WITH COMPOUND LEAVES

Rattlebush

Sesbania drummondii (Rydb.) Cory
[*Daubentonia drummondii* Rydb.]
Legume Family
Leguminosae

Description: Broad-leaved deciduous shrub, up to 10 feet tall, with upper branches usually dying back in winter and only lower plant parts woody; young shoots whitish silky-hairy; compound leaves (up to 8 inches long) composed of fifteen to twenty somewhat oblong leaflets (up to 1½ inches long and about ⅕ inch wide), leaf stalks usually hairy, alternately arranged; yellow pealike flowers (½–⅔ inch long), often marked with reddish lines and fine spots, borne in clusters of ten to thirty (racemes) in leaf axils; four-winged fruit pod (about 2 inches long and less than ½ inch wide). (*Note:* Dried seeds are loose, so mature pods rattle when shaken, giving the plant its common name.)

Flowering period: June through September.

Habitat: Brackish marshes and sandy beaches; moist or wet sands, waste areas, and forest edges.

Wetland indicator status: FACW.

Range: Northwestern Florida (Panhandle) west along the Coastal Plain to Texas.

Similar species: Hemp Sesbania or Coffeeweed (*S. exaltata*, formerly *S. macrocarpon*) is an annual herb with two to six yellow flowers per raceme, and its pods are longer (up to 8 inches long), narrower (less than ⅕ inch wide), and lack wings; it occurs from southwestern New York to Florida and west to Texas; it is FACW−. Gray Nicker or Nickerbean (*Caesalpinia bonduc*, formerly *C. crista*) occurs as a prickly vinelike tropical shrub along the borders of mangrove swamps; its compound leaves are very long (up to 2 feet) and divided into many pairs of leaflets and its fruit pods are brown and densely prickled; it is FACU+.

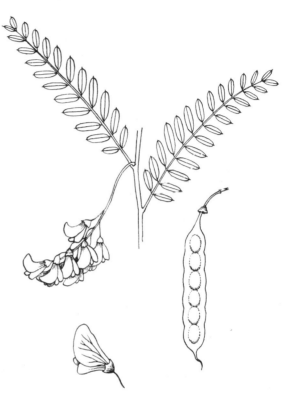

Rattlebush

DECIDUOUS SHRUBS WITH SIMPLE, ENTIRE, ALTERNATE LEAVES

Saltwater False Willow or Narrow-leaved Groundsel Bush

Baccharis angustifolia Michx.

Composite or Aster Family
Compositae (*Asteraceae*)

Description: Narrow-leaved, low deciduous shrub (up to 7 feet tall); simple, entire (or few-toothed), linear to narrowly egg-shaped, somewhat leathery or fleshy, shiny leaves (up to 3 inches long and to $\frac{1}{5}$ inch wide), stalkless, lower leaves often toothed, alternately arranged; many small yellow disk flowers in stalkless or short-stalked heads borne singly or in clusters, separate male and female flowers borne on separate plants (dioecious); shiny, many-ribbed nutlets covered with long white hairs.

Flowering period: September through October.

Habitat: Upper edges of salt and brackish marshes, mangrove swamps, and coastal hammocks.

Wetland indicator status: FACW+.

Range: North Carolina along the Coastal Plain to southern Florida, west to Texas.

Similar species: Groundsel Bush (*B. halimifolia*) has toothed, broad leaves; it is FAC.

Sea Grape

Coccoloba uvifera (L.) L.
Buckwheat Family
Polygonaceae

Description: Broad-leaved deciduous tropical shrub or tree, up to 50 feet tall, usually less than 25 feet; bark smooth and brown; simple, entire, somewhat roundish, leathery leaves (about 3–10 inches wide) with heart-shaped bases, short-stalked and sheathed, reddish veins, stalks, and sheaths, new leaves often bronze-red, alternately arranged; many small creamy whitish, five-lobed flowers (about $\frac{1}{4}$ inch wide) borne on short stalks in many rows on narrow, spikelike drooping clusters (racemes, 3–14 inches long); purplish, fleshy, edible grapelike fruits ($\frac{1}{4}$–$\frac{3}{4}$ inch wide) borne in drooping clusters.

Flowering period: Spring and summer.

Habitat: Salt flats, sandy margins of salt marshes and mangrove swamps, upper portions of sandy beaches; sand dunes and coastal thickets; also widely planted for landscaping.

Wetland indicator status: FAC.

Range: Peninsular Florida; also in Bermuda, the Bahamas, the Caribbean, and elsewhere in tropical America.

SEE ALSO Corkwood (*Leitneria floridana*).

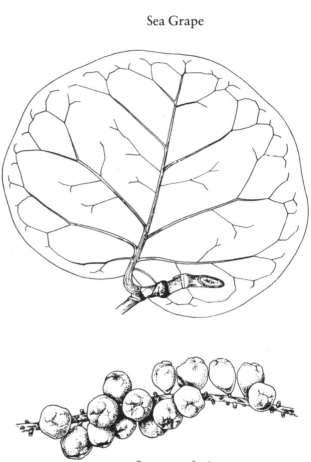

Sea Grape

Saltwater False Willow

Sea grape fruits

DECIDUOUS SHRUBS WITH SIMPLE, TOOTHED, ALTERNATE LEAVES

Groundsel Bush or Sea Myrtle

Baccharis halimifolia L.
Composite or Aster Family
Compositae (Asteraceae)

Description: Broad-leaved deciduous shrub, up to 10 feet tall; simple, thick, egg-shaped leaves (up to 2½ inches long), mostly coarsely toothed above middle of leaf, uppermost leaves entire, alternately arranged; white flowers in small heads in mostly stalked clusters forming terminal leafy inflorescences, male and female flowers borne on separate plants (dioecious); nutlet with whitish hairy bristles appearing cottony.

Flowering period: August into November.

Habitat: Irregularly flooded salt, brackish, and tidal fresh marshes; nontidal swamps, hydric hammocks, and open woods and thickets along the coast.

Wetland indicator status: FAC.

Range: Massachusetts south along the Coastal Plain and in the Piedmont to Florida, west to Arkansas and Texas.

Similar species: Groundsel Tree (*B. glomeruliflora*) occurs from South Carolina south; its flowering heads are mostly not stalked; it is FACW. Saltwater False Willow (*B. angustifolia*) has linear to narrowly egg-shaped, mostly entire leaves; it is FACW+. Broom-bush False Willow (*B. dioica*) occurs in Florida, the Bahamas, and the Caribbean; its egg-shaped to somewhat lance-shaped leaves are entire; it is FACW.

Groundsel Bush

EVERGREEN TREES WITH FANLIKE LEAVES

Cabbage Palm

Sabal palmetto (Walter) Lodd. ex J. A.
   & J. H. Schultes

Palm Family
Arecaceae

Description: Evergreen palm tree, up to 75 feet tall; trunk unbranched, bearing fanlike leaves at top; bark rough, grayish to brownish; stiff, green, compound, fanlike leaves (up to 7 feet wide and long) composed of numerous long-pointed, drooping leaflets with prominent midrib and threadlike fibers at margins; small fragrant, white, six-"petaled" tubular flowers (about ⅕ inch long) borne in many drooping, branched clusters; shiny, black, berrylike, one-seeded fruits (less than ½ inch wide).

Flowering period: June and July.

Habitat: Brackish marshes, upper edges of salt marshes, mangrove swamps, and tidal swamps; nontidal hydric hammocks, wet prairies, and maritime forests; also widely planted for landscaping.

Wetland indicator status: FAC.

Range: Southeastern North Carolina along the Coastal Plain to Florida, west to Alabama.

Similar species: Bluestem or Dwarf Palmetto (*S. minor*) is a common shrub of freshwater wetlands in the Southeast; it usually lacks a stem, and its bluish green leaves lack a prominent midrib; it is FACW. Saw Palmetto (*Serenoa repens*) occurs along the upper edges of salt marshes but is more common on uplands; it is a shrub with fanlike leaves and leaf stalks armed with sawlike, spiny teeth; it is FACU.

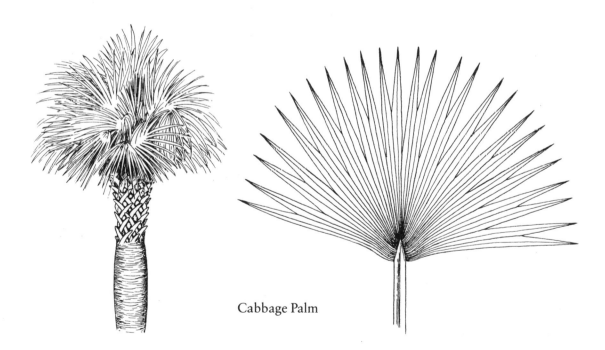

Cabbage Palm

EVERGREEN TREES WITH SIMPLE ENTIRE LEAVES

Buttonwood

Conocarpus erectus L.

White Mangrove Family
Combretaceae (*Terminaliaceae*)

Description: Broad-leaved evergreen tropical shrub or tree, up to 60 feet or taller; older bark ridged and flaky; twigs reddish brown and angled or winged; simple, entire, leathery evergreen leaves (about 1–4 inches long and to about 1¾ inches wide) with sharp tips and tapered bases, short-stalked, pair of salt glands at base, alternately arranged; small greenish, five-lobed, hairy flowers in heads (about ⅓ inch wide) borne in clusters (racemes, about 1–2 inches long); pinecone-like, roundish fruit capsule (about ⅓ inch long) with many overlapping scales.

Flowering period: Yearlong.

Habitat: Upper edges of mangrove swamps and salt flats; hydric hammocks.

Wetland indicator status: FACW+.

Range: Peninsular Florida; also in Bermuda, the West Indies, tropical America, and Africa.

White Mangrove

Laguncularia racemosa (L.) C. F. Gaertn.

White Mangrove Family
Combretaceae (*Terminaliaceae*)

Description: Broad-leaved evergreen tropical shrub or tree, up to 70 feet high or taller; above-ground pneumatophores usually absent but sometimes present; older bark grayish with vertical ridges and furrows; young bark light brown and smooth; twigs light reddish brown and thickened at nodes; simple, entire, leathery or fleshy, thick, yellow-green, oval-shaped evergreen leaves (about 1–4 inches long and 2 inches wide) with somewhat rounded tips, usually shallow-notched, stalked, paired salt glands on green (sometimes purplish) leaf stalk, oppositely arranged; few to many small, greenish white, five-petaled flowers borne in clusters on spikes (about 1–2½ inches long); leathery, roundish, reddish fruit capsule (less than 1 inch long).

Flowering period: Yearlong, but mostly May and June.

Habitat: Mangrove swamps (usually irregularly flooded zone).

Wetland indicator status: FACW+.

Range: Peninsular Florida; also in Bermuda, the Bahamas, and elsewhere in the Caribbean and tropical America.

White Mangrove

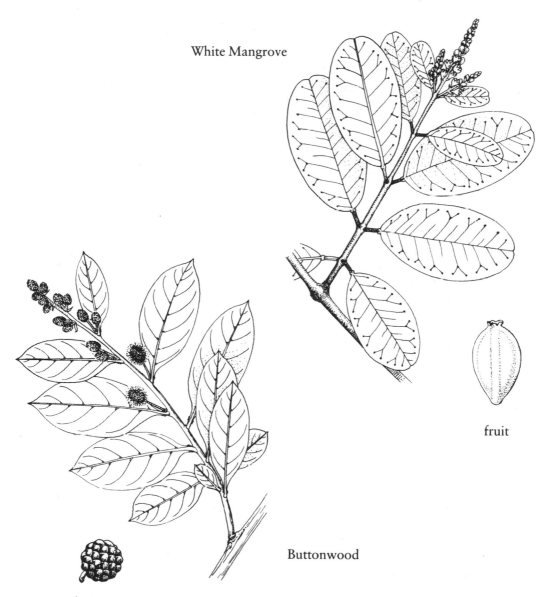

fruit

Buttonwood

fruit

Red Mangrove

Rhizophora mangle L.

Mangrove Family
Rhizophoraceae

Description: Broad-leaved evergreen tropical shrub or tree, up to 80 feet tall, with conspicuous arching prop or stilt roots; older bark reddish and smooth, young bark grayish; twigs silverish to shiny dark brown; simple, entire, smooth, leathery leaves (2–6½ inches long and to 3 inches wide) with prominent midrib, dark green and shiny above, lighter green (often speckled) below, stalked, oppositely arranged; pale yellowish or cream-colored, somewhat leathery, four-petaled flowers borne on stalks (about ½–1½ inches long) in clusters of two or more; elongate greenish fruit capsule (about 1 inch long before germination), seed germinates while on plant giving fruit a curved, elongate appearance.

Flowering period: Yearlong but mostly in spring and early summer.

Habitat: Mangrove swamps, salt marshes, and fresh marshes (near coast) of the Florida Everglades.

Wetland indicator status: OBL.

Range: Northeastern Florida south along Florida coast to Cedar Key; also in Bermuda, the West Indies, tropical America, Africa, and Pacific islands.

Black Mangrove

Avicennia germinans (L.) L.
[*Avicennia nitida* Jacq.]

Vervain Family
Verbenaceae

Description: Broad-leaved evergreen tropical shrub or tree, up to about 80 feet tall; numerous erect pneumatophores growing aboveground from roots; dark, scaly bark with orange-red inner bark, older bark gray to blackish; twigs with distinct nodes; simple, entire, leathery, smooth evergreen leaves (1–3¼ inches long), shiny green above, fine-hairy grayish below, lateral veins conspicuous, short-stalked, oppositely arranged; numerous small white, four-"petaled" tubular flowers (about ½ inch wide) borne in stalked clusters (panicles, about 1–2 inches long); light green, flattened, slightly hairy fruit capsule (about 1–2 inches long).

Flowering period: Spring and early summer.

Habitat: Mangrove swamps, salt marshes, salt flats, and shallow estuarine water.

Wetland indicator status: OBL.

Range: St. Augustine, Florida, to Texas; also in Bermuda, the Bahamas, the West Indies, tropical America, and west coast of Africa.

Similar species: Although not similar except for being an evergreen tropical shrub, Brazilian Pepper (*Schinus terebinthifolius*), an invasive exotic, occurs along the borders of mangrove swamps; it has compound leaves of three to eleven leaflets and bears many bright red berries; it is FAC.

Live Oak

Quercus virginiana Mill.

Beech Family
Fagaceae

Description: Broad-leaved evergreen tree, up to 60 feet tall, with wide-spreading branches; bark dark to reddish brown, ridged and slightly furrowed (sometimes blocky); simple, mostly entire (some with few teeth), thick, evergreen, oblong to narrowly egg-shaped leaves (to 5 inches long and to 2½ inches wide), dark green and shiny above, gray and often hairy below, alternately arranged; somewhat narrow, elongate acorns (about 1 inch long) with cone-shaped (turbinate) caps, borne on stalks singly or in clusters of two to five.

Flowering period: March and April.

Habitat: Edges of salt, brackish, and tidal fresh marshes, and tidal swamps; nontidal forested wetlands; hydric hammocks, sand dunes, maritime forests, and well-drained uplands.

Wetland indicator status: FACU+.

Range: Southeastern Virginia along the Coastal Plain to Florida, west to Texas and Oklahoma; also in Mexico.

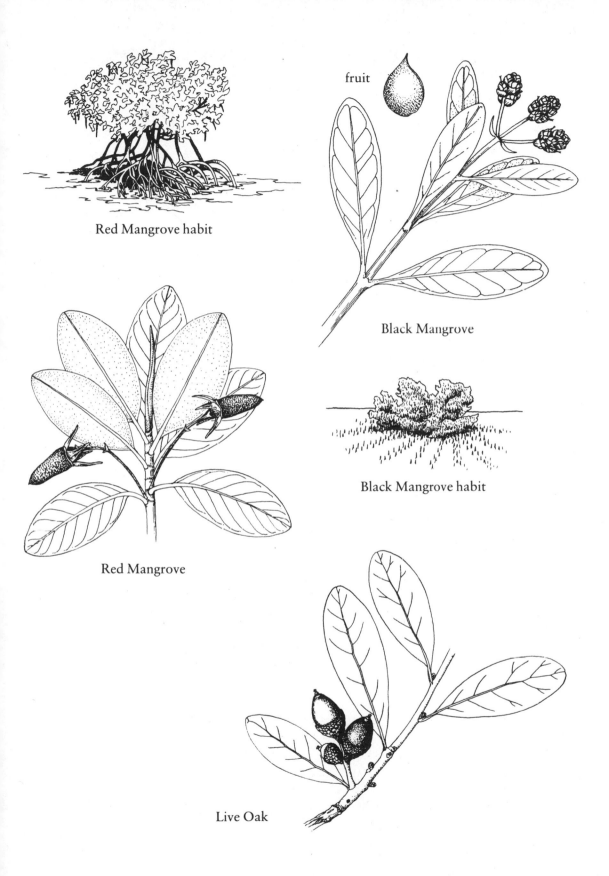

fruit

Red Mangrove habit

Black Mangrove

Black Mangrove habit

Red Mangrove

Live Oak

Deer Pea or Hairy-pod Cow Pea

Vigna luteola (Jacq.) Benth.
[*Vigna repens* (L.) Kuntze]

Legume Family
Leguminosae

Description: Perennial, climbing or trailing herbaceous vine with usually hairy stems up to 10 feet long; compound leaves divided into three entire, lance-shaped or linear or egg-shaped leaflets (up to 3¼ inches long and 1½ inches wide) sparsely covered by short hairs, hairy leaf stalks, alternately arranged; small yellow, pealike flowers (less than 1 inch long) borne on long-stalked inflorescences (racemes) in leaf axils, hairy flower stalks; elongate, flattened, short-haired fruit pod (up to 2¾ inches long).

Flowering period: July through September.

Habitat: Borders of salt, brackish, and tidal fresh marshes and mangrove swamps; ditches, swales, wet sands, flatwoods, low fields, and waste places.

Wetland indicator status: FACW.

Range: Southeastern North Carolina to Florida, west to southeastern Texas; also in tropical America.

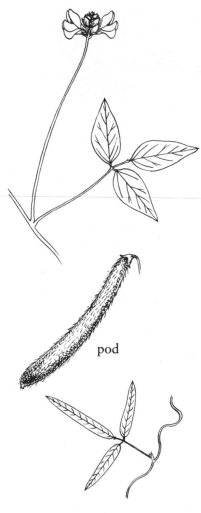

pod

Deer Pea

HERBACEOUS VINES WITH SIMPLE LEAVES

Climbing Milkweed or Vine Milkweed

Cynanchum angustifolium Pers.
[*Cynanchum palustre* (Pursh) Heller]

Milkweed Family
Asclepiadaceae

Description: Slender, perennial, twining herbaceous vine with milky sap; simple, entire, narrow linear leaves (up to 3⅓ inches long and less than ⅕ inch wide), stalkless, oppositely arranged; several small greenish white flowers (less than ⅕ inch wide), sometimes red- or purple-tinged, borne on stalks in axillary clusters (umbels, over ⅓ inch wide); elongate, tapering fruit pod (follicle, up to 2½ inches long and about ¼ inch wide).

Flowering period: June through July.

Habitat: Upper edges of salt and brackish marshes; shell mounds, spoil banks and flats, and moist soils of coastal hammocks.

Wetland indicator status: FACW.

Range: North Carolina to Florida Keys, west to Texas; also in the Bahamas and the West Indies.

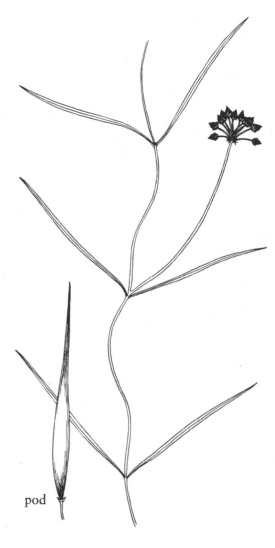

pod

Climbing Milkweed

Salt Marsh Morning Glory

Ipomoea sagittata Poir.

Morning Glory Family
Convolvulaceae

Description: Perennial, trailing or twining herbaceous vine; stems smooth; simple, entire, three-lobed, narrow arrowhead-shaped leaves (up to 3¾ inches long and to ¾ inch wide), stalked (up to 1¼ inches long), alternately arranged; large rose to lavender (rarely white), funnel-shaped flowers (to 3¾ inches wide) with reddish purple centers, borne singly on stalks and subtended by thick, somewhat leathery sepals; somewhat rounded fruit capsule (about ½ inch wide) with a short projection at the top.

Flowering period: April through September.

Habitat: Edges of brackish and tidal fresh marshes; interdunal swales, nontidal marshes along the coast, and moist sandy roadsides.

Wetland indicator status: FACW.

Range: North Carolina south along the Coastal Plain to Florida, west to Texas; also in the West Indies.

Similar species: Hedge Bindweed (*Calystegia sepium*, formerly *Convolvulus sepium*) has broader triangle-shaped leaves, and its flowers are white to pinkish purple; it is FAC. Beach Morning Glory (*Ipomoea stolonifera*) has been reported in tidal fresh marshes, although it is more typically a dune plant; it is a trailing vine, rooting at the nodes, and has white flowers with yellow or purplish centers; it is FACU. Common Morning Glory (*I. purpurea*) has hairy stems and hairy leaflike sepals subtending the purplish, pinkish, or whitish flowers; it is FACU. Red Morning Glory (*I. coccinea*) has red flowers; it is FAC.

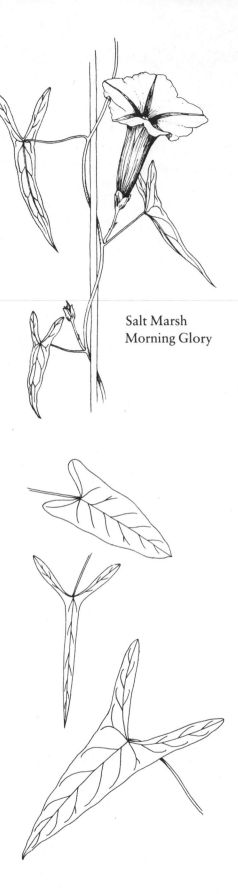

Salt Marsh
Morning Glory

WOODY VINES

Mangrove Swamp Vine

Rhabdadenia biflora (Jacq.) Muell. Arg.

Dogbane Family
Apocynaceae

Description: Woody tropical vine; simple, entire, somewhat fleshy leaves (up to 3½ inches long) with sharp-pointed tips (mucronate), thin-stalked, oppositely arranged; fragrant, whitish, five-lobed tubular flowers (about 2 inches long and to 1¾ inches wide) borne singly, in pairs, or on long-stalked clusters (racemes or cymes); linear fruit pod (follicle, about 6 inches long and about ⅕ inch wide).

Flowering period: Yearlong.

Habitat: Mangrove swamps.

Wetland indicator status: FACW+.

Range: Peninsular Florida; also in the Bahamas and the Caribbean.

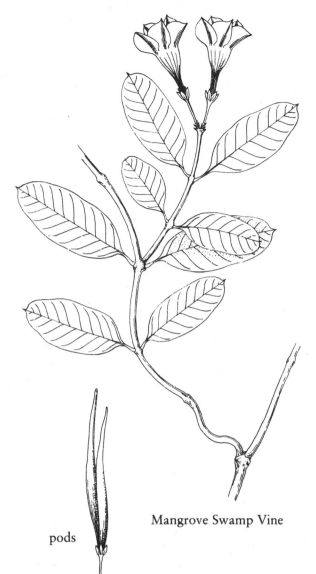

pods

Mangrove Swamp Vine

SEE ALSO Ear-leaf Greenbrier (*Smilax auriculata*), described under Laurel-leaved Greenbrier (*S. laurifolia*), and Coin-vine (*Dalbergia ecastophyllum*).

*Plants of Tidal Fresh
Coastal Wetlands*

FERNS

Cinnamon Fern

Osmunda cinnamomea L.

Royal Fern Family
Osmundaceae

Description: Erect fern up to 5 feet tall with cinnamon-colored, woolly stalks; two types of fronds—(1) sterile, leaflike fronds, compound blades (up to 12 inches wide) with up to twenty-five pairs of leaflets, alternately arranged or nearly oppositely arranged, each leaflet with a tuft of brownish hairs at base, and (2) fertile fronds bearing compound "leaflets" (up to 1½ inches long) with sporangia first greenish and quickly becoming cinnamon brown. (*Note:* Fertile fronds are surrounded by sterile fronds.)

Fruiting period: March through May.

Habitat: Tidal swamps; nontidal forested wetlands, hydric hammocks, stream banks, seepage slopes, margins of bogs, and wet rock ledges; subacid soils.

Wetland indicator status: FACW+.

Range: Labrador to Minnesota, south to Florida, Texas, and New Mexico.

Similar species: Virginia Chain Fern (*Woodwardia virginica*) has shiny, dark purplish brown stalks and only one type of frond (leaflike); it is OBL.

Royal Fern

Osmunda regalis L. var. *spectabilis* (Willd.) Gray

Royal Fern Family
Osmundaceae

Description: Medium-height to tall, erect fern, 1½–6 feet high, forming tussocks or clumps; rhizomes black and wiry; stalk smooth, straw-colored, and reddish at base; compound leaves (fronds, up to 22 inches wide) twice divided into separate, oblong, short-stalked leaflets in five to eleven pairs per branchlet; fertile leaves with light brown spore-bearing leaflets at the top forming a terminal inflorescence (panicle, up to 12 inches long); fruits sporangia.

Fruiting period: March through June.

Habitat: Tidal fresh marshes and swamps; nontidal marshes, swamps, wet meadows, and moist woods.

Wetland indicator status: OBL.

Range: Newfoundland to Saskatchewan, south to Florida and Texas.

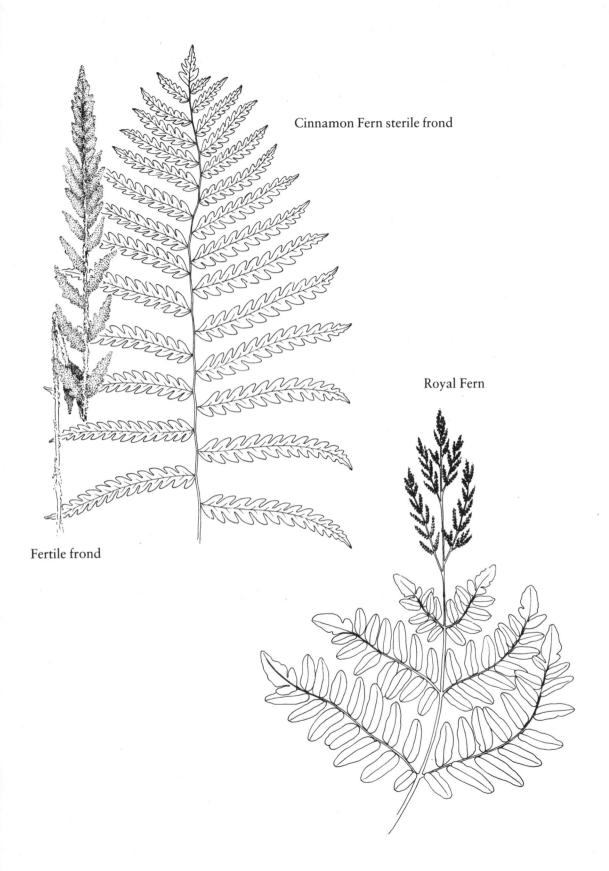

Cinnamon Fern sterile frond

Royal Fern

Fertile frond

Sensitive Fern

Onoclea sensibilis L.

Polypody Fern Family
Polypodiaceae

Description: Low to medium-height, erect fern, up to 3½ feet tall; rhizomes brown, usually smooth, and creeping near surface with fibrous rootlets; stalk smooth, thickened at base, yellow with brown; compound light green leaves (up to 14 inches long and 16 inches wide) divided into shallowly lobed leaflets, uppermost connected to one another along stalk, lower leaflets separate; separate fertile frond arising from rhizome, bearing beadlike fertile leaflets that become dark brown at maturity; sporangia.

Fruiting period: May into September.

Habitat: Tidal fresh marshes; nontidal marshes, meadows, swamps, and moist woodlands.

Wetland indicator status: FACW.

Range: Newfoundland to Ontario, Minnesota, and South Dakota, south to Florida and Texas.

Marsh Fern

Thelypteris thelypteroides (Michx.)
   J. Holub
[*Thelypteris palustris* Schott]

Polypody Fern Family
Polypodiaceae

Description: Medium-height, erect fern, up to 28 inches tall; rhizomes black and branched; stalks about 9 inches long, smooth, slender, and pale green above and black at base; compound light green or yellow-green leaves (up to 16 inches long and 8½ inches wide) divided into twelve or more pairs of lance-shaped leaflets with rounded ends, two types of leaves (fertile and sterile), fertile leaves more erect and on longer stalks than sterile leaves; fruit dots (sori) borne on undersides of upper leaflets near midvein.

Fruiting period: June through September.

Habitat: Tidal fresh marshes, occasionally along upper edges of salt and brackish marshes; nontidal marshes, shrub swamps, and forested wetlands.

Wetland indicator status: FACW+.

Range: Newfoundland to Ontario and Manitoba, south to Florida and Texas.

Similar species: New York Fern (*T. noveboracensis*) has greatly tapered lower leaflets and occurs from northeastern North Carolina north on the Coastal Plain; it is FAC+.

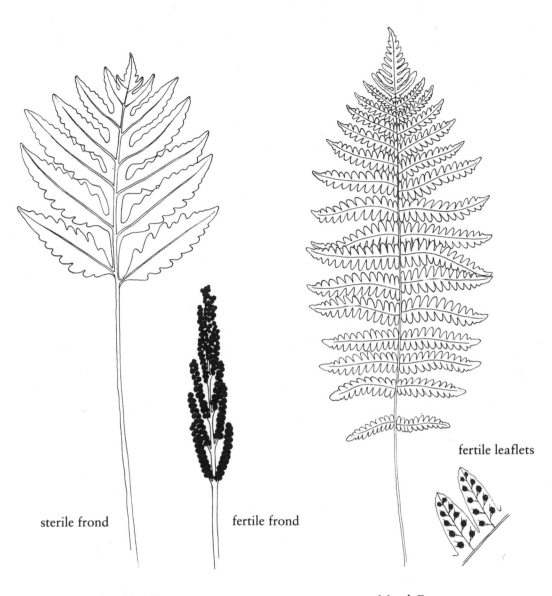

sterile frond   fertile frond

Sensitive Fern

fertile leaflets

Marsh Fern

Net-veined Chain Fern

Woodwardia areolata (L.) Moore

Polypody Fern Family
Polypodiaceae

Description: Erect fern, up to 2½ feet tall; two types of fronds—(1) sterile fronds, appearing like compound leaves but actually mostly deeply lobed, forming numerous leafletlike lobes, blades up to 7 inches wide, leaf margins finely toothed, and (2) fertile fronds with narrow, elongate leaflets bearing chainlike rows of oblong sori (containing spores).

Fruiting period: June through September.

Habitat: Tidal swamps; nontidal forested wetlands (especially on Coastal Plain), margins of bogs, and seepage slopes in highly acidic soils.

Wetland indicator status: OBL.

Range: Nova Scotia south to Florida and Texas (mainly along the coast), west to Michigan and Missouri.

Similar species: Sensitive Fern (*Onoclea sensibilis*) has entire leaf margins and spores enclosed in beadlike structures arranged in rows on fertile frond; it is FACW.

Virginia Chain Fern

Woodwardia virginica (L.) Smith

Polypody Fern Family
Polypodiaceae

Description: Erect fern, up to 4 feet tall, with shiny, dark purplish brown stalks; one type of frond (leaflike), compound blades (up to 12 inches wide) with many-lobed leaflets alternately arranged along stalk, middle leaflets longest; oblong, spore-bearing sori borne on underside of fertile leaflets in two rows (one on each side of midrib).

Fruiting period: June through September.

Habitat: Tidal swamps; nontidal forested wetlands, hydric hammocks, shrub wetlands, and bogs in moderately or highly acidic soils.

Wetland indicator status: OBL.

Range: Nova Scotia, Ontario, and Illinois south to Florida and Texas.

Similar species: Resembles Cinnamon Fern (*Osmunda cinnamomea*), which has cinnamon-colored woolly stalks and two types of fronds (sterile and fertile); it is FACW+.

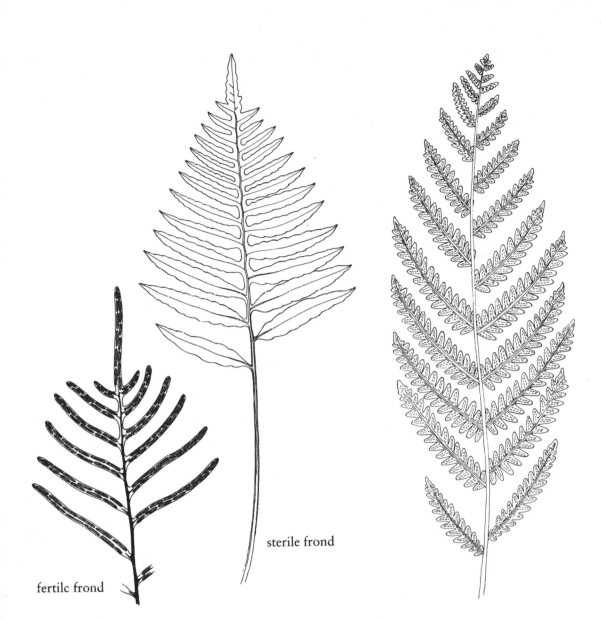

fertile frond

sterile frond

Net-veined Chain Fern

Virginia Chain Fern

GRASSES

Lowland Broomsedge
or Bushy Beardgrass

Andropogon glomeratus (Walter) B.S.P.

Grass Family
Gramineae (*Poaceae*)

Description: Medium-height to tall perennial grass, up to 5 feet high, occurring in dense clumps; linear leaves (up to 12 inches long) with broad overlapping sheaths; densely branched, feathery inflorescence (up to 2 inches long) with hairlike spikelets forming between upper leaves.

Flowering period: August through October.

Habitat: Tidal fresh marshes and upper edges of salt marshes; nontidal marshes, wet soils, and sandy grounds along the coast.

Wetland indicator status: FACW+.

Range: Maine to Ohio, south to Florida and Texas; also in western states.

Slender Spikegrass

Chasmanthium laxum (L.) H. Yates
[*Uniola laxa* (L.) B.S.P.]

Grass Family
Gramineae (*Poaceae*)

Description: Medium-height perennial grass, up to 4 feet tall; slender stems; elongate, flat, sometimes inwardly rolled leaves with a fine narrow tip; narrow terminal inflorescence (up to 18 inches long) with short ascending branches bearing spikelets.

Flowering period: June through October.

Habitat: Tidal swamps; nontidal forested wetlands and moist woods and fields (mainly Coastal Plain).

Wetland indicator status: FACW−.

Range: Long Island, New York, to Kentucky and Oklahoma, south to Florida and Texas.

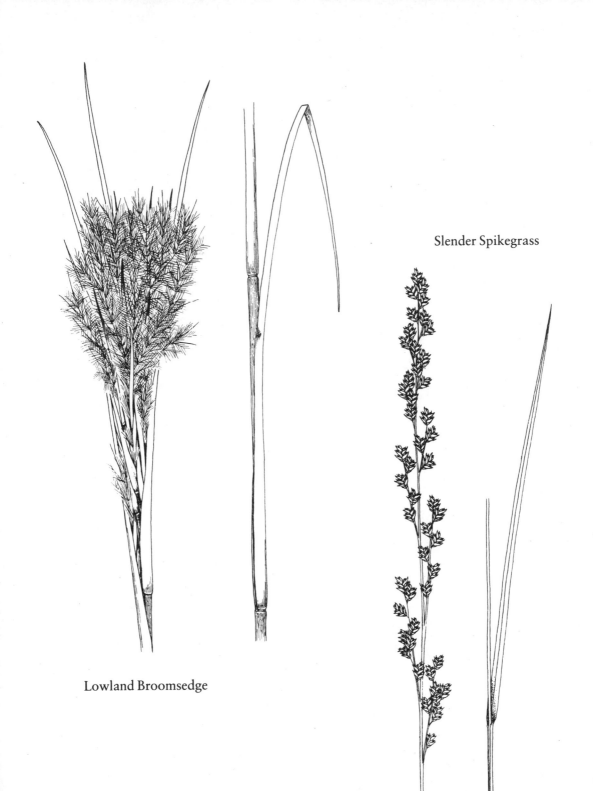

Slender Spikegrass

Lowland Broomsedge

Wood Reed

Cinna arundinacea L.

Grass Family
Gramineae (Poaceae)

Description: Medium-height to tall perennial grass, up to 5 feet high; stem with five to ten nodes; ligule membranous; slightly rough-margined, flat leaves (up to 16 inches long and ½ inch wide), surface often rough; narrow terminal inflorescence (up to 12 inches long) with dense, somewhat drooping ascending branches.

Flowering period: August through October.

Habitat: Tidal swamps; nontidal forested wetlands and moist woods.

Wetland indicator status: FACW.

Range: Maine to Ontario and South Dakota, south to Georgia and Texas.

Walter Millet

Echinochloa walteri (Pursh) A. Heller

Grass Family
Gramineae (Poaceae)

Description: Medium-height to tall, erect annual grass, 3½–6½ feet high; stems round, hollow, and erect; long, tapering leaves (up to 20 inches long and 1 inch wide), leaf sheaths short, coarse-hairy; dense terminal inflorescence (panicle, 4–12 inches long) bearing numerous erect spikes with many spikelets covered by long bristles (awns, often 1¼ inches long).

Flowering period: June through October.

Habitat: Tidal fresh marshes; nontidal fresh and alkaline marshes, swamps, and shallow waters.

Wetland indicator status: OBL.

Range: Massachusetts to Florida and Texas; inland, New York to Wisconsin and Minnesota.

Similar species: Barnyard Grass (*E. crusgalli*) is usually less than 3½ feet tall, and its spikelets typically lack bristles (awns); it is FACW−.

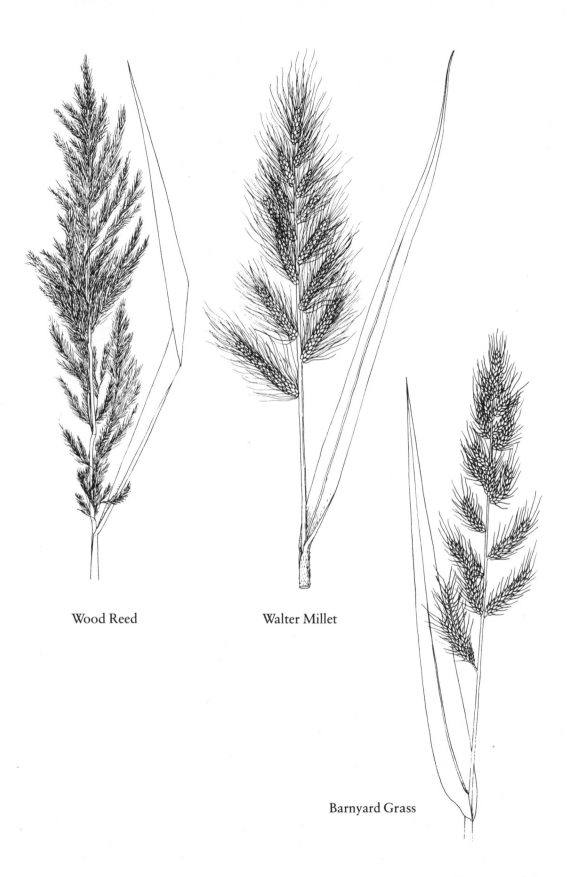

Wood Reed

Walter Millet

Barnyard Grass

Virginia Rye Grass

Elymus virginicus L.

Grass Family
Gramineae (*Poaceae*)

Description: Medium-height to tall, erect perennial grass, 1½–4½ feet high; stems stout and forming clumps; leaf sheaths smooth, leaf blades (⅕–⅘ inch wide) usually rolled inwardly; terminal, unbranched inflorescence (panicle, up to 6 inches long) crowded with spikelets (glumes and lemmas) having conspicuously long bristles (awns).

Flowering period: June through October.

Habitat: Tidal fresh marshes and borders of brackish marshes; moist woods, meadows, thickets, shores, and prairies.

Wetland indicator status: FAC.

Range: Newfoundland to Alberta, south to Florida and Arizona.

Similar species: Walter Millet (*Echinochloa walteri*) also has spikelets with long bristles (awns), but its panicle is branched; it is OBL.

Giant Plume Grass or Beard Grass

Erianthus giganteus (Walter) F. T. Hubb
    non Muhl.

Grass Family
Gramineae (*Poaceae*)

Description: Tall perennial grass, up to 14 feet high, growing in clumps; stems smooth or hairy, often hairy below inflorescence, nodes with long bristly hairs, especially when young; long linear leaves (up to 20 inches long and about 1 inch wide) with smooth or hairy surfaces, rough margins, and smooth, rough, or long-hairy leaf sheaths; inconspicuous flowers borne in purplish or silvery, dense terminal inflorescence (panicle, up to 16 inches long and to 6 inches wide) with ascending branches, spikelets long-hairy and rough with elongate rough bristles (awns).

Flowering period: September through October.

Habitat: Slightly brackish and tidal fresh marshes; nontidal marshes, ditches, wet swales, savannahs, moist open areas, and edges of swamps.

Wetland indicator status: FACW.

Range: New York south, generally along the Coastal Plain to Florida and Texas; also in Tennessee and Kentucky.

Similar species: Common Reed (*Phragmites australis*) and Giant Reed (*Arundo donax*) are not as hairy in general appearance, and their leaf sheaths are not hairy, although the former species has hairy ligules; both are FACW.

Rice Cutgrass

Leersia oryzoides (L.) Swartz

Grass Family
Gramineae (*Poaceae*)

Description: Medium-height to tall, erect perennial grass, 2–5 feet high; stems rough-hairy, erect or lying flat on ground at base, then ascending; tapered yellowish green leaves (up to 8 inches long and ½ inch wide), very rough margins with stiff hairs, leaf sheaths rough-edged; open terminal inflorescence (panicle, 4–8 inches long) with spreading or ascending slender branches, spikelets (up to ½ inch long) arising from upper half or two-thirds of branches.

Flowering period: June into October.

Habitat: Tidal fresh marshes, occasionally slightly brackish marshes; nontidal swamps, wet meadows, marshes, ditches, and muddy shores.

Wetland indicator status: OBL.

Range: Quebec and Nova Scotia to eastern Washington, south to Florida, New Mexico, and California.

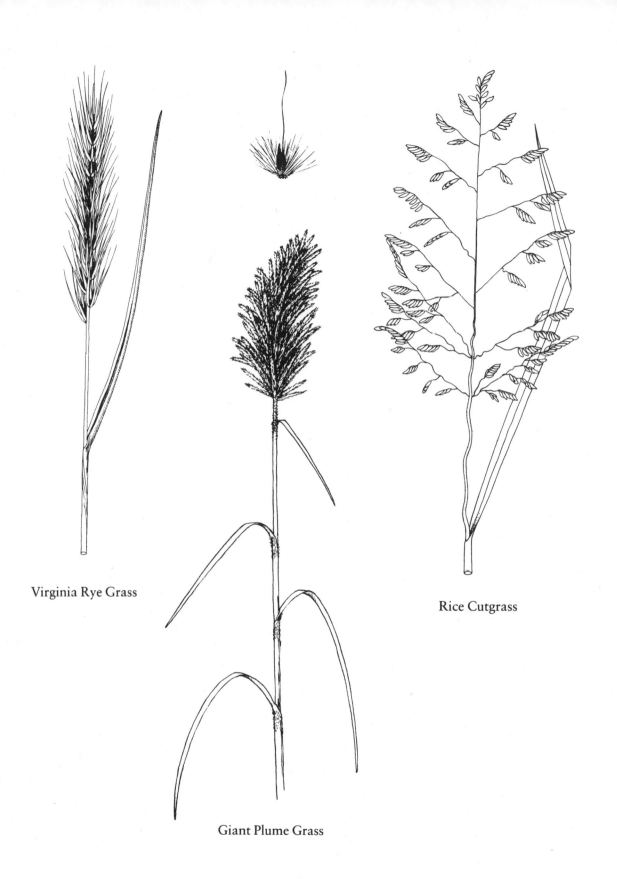

Virginia Rye Grass

Giant Plume Grass

Rice Cutgrass

Gulf Muhly

Muhlenbergia capillaris (Lam.) Trin.

Grass Family
Gramineae (*Poaceae*)

Description: Medium-height perennial grass, up to 4 feet tall, occurring in clumps, stems slender and smooth; mostly basal linear leaves (up to 16 inches long and about ⅕ inch wide) with rough surfaces, margins, and leaf sheaths; inconspicuous, rough, purplish flowers borne on thin, hairlike, rough branches in open, spreading terminal inflorescence (panicle, up to 20 inches long and to 8 inches wide); purplish seed.

Flowering period: September through October.

Habitat: Tidal fresh marshes, upper edges of salt and brackish marshes; interdunal swales, nontidal marshes, upland forests, savannahs, and pinelands.

Wetland indicator status: FACU.

Range: Massachusetts and Kansas to Florida and Texas; also in Mexico and the West Indies.

Maidencane or Paille Fine

Panicum hemitomon J. A. Schultes

Grass Family
Gramineae (*Poaceae*)

Description: Medium-height to tall perennial grass, up to 8 feet high, often forming extensive colonies in shallow water; rhizomes dense, elongate, often rooting at nodes; tapered linear leaves (up to 12 inches long and ⅗ inch wide) with rough upper surfaces and margins, leaf sheaths overlapping and sometimes hairy, ligule hairy; narrow terminal inflorescence (panicle, up to 10 inches long and to ⅝ inch wide) composed of three- to five-nerved spikelets on rough, appressed branches.

Flowering period: June and July.

Habitat: Slightly brackish and tidal fresh marshes; nontidal marshes, shallow waters of lakes and ponds, and ditches.

Wetland indicator status: OBL.

Range: New Jersey south to Florida, west to Texas; also in Tennessee and South America.

American Cupscale or Bagscale

Sacciolepis striata (L.) Nash

Grass Family
Gramineae (*Poaceae*)

Description: Medium-height perennial grass, 20–40 inches tall, sometimes forming dense stands; stolons creeping; stems often reclining and rooting at nodes; flat, linear leaves (up to 8 inches long and to ½ inch wide) with somewhat heart-shaped bases, smooth surfaces, rough-hairy margins, upper leaves often downward-pointing, ligule membranous; narrow, cylindrical, terminal spikelike inflorescence (panicle, up to 10 inches long and about ½ inch wide) bearing many stalked, many-nerved (ribbed) spikelets (about ⅕ inch long).

Flowering period: July through October.

Habitat: Water's edge of tidal fresh marshes; nontidal marshes, open water, and ditches.

Wetland indicator status: OBL.

Range: Southern New Jersey along the Coastal Plain to Florida, west to Texas; also in Tennessee, Oklahoma, and the West Indies.

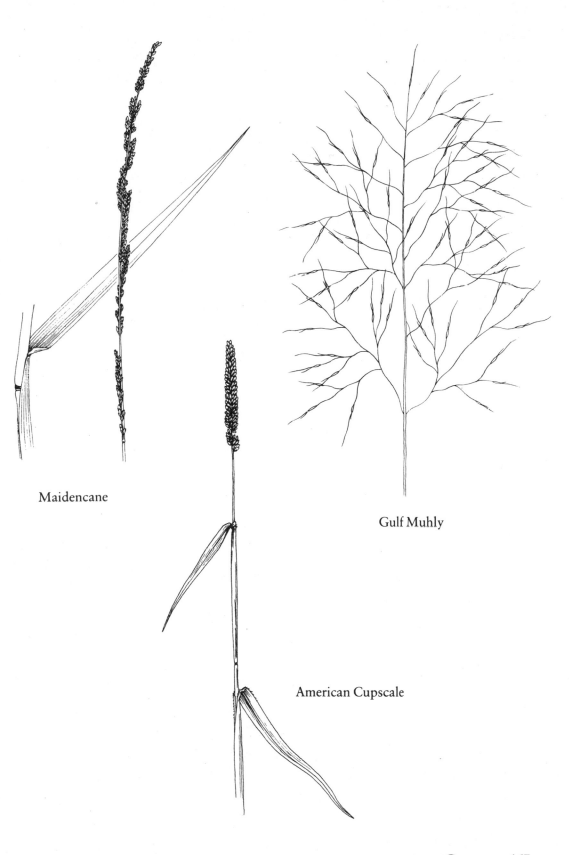

Maidencane

Gulf Muhly

American Cupscale

Wild Rice

Zizania aquatica L.

Grass Family
Gramineae (Poaceae)

Description: Tall, erect annual grass, up to 10 feet high; stems stout, sometimes lying flat on ground at base, then ascending; large, soft, flat, tapered leaves (up to 48 inches long and 2 inches wide) with rough margins; open terminal inflorescence (panicle, 4–24 inches long) divided into two parts, lower branches bearing drooping male spikelets (branches first erect, later open-spreading) and upper branches bearing female spikelets (branches first appressed, then ascending after fertilization).

Flowering period: May into October.

Habitat: Tidal fresh marshes and slightly brackish marshes (regularly and irregularly flooded zones); stream borders, shallow waters, and nontidal marshes.

Wetland indicator status: OBL.

Range: Eastern Quebec and Nova Scotia to Manitoba, south to Florida and Louisiana; also in Arizona.

Similar species: Leaves resemble those of Giant Cutgrass or Southern Wild Rice (*Zizaniopsis miliacea*), but its inflorescence is not divided into separate sections containing male and female parts; instead, each branch is so divided; it is OBL.

Giant Cutgrass
or Southern Wild Rice

Zizaniopsis miliacea (Michx.) Doell & Aschers.

Grass Family
Gramineae (Poaceae)

Description: Tall perennial grass, up to 10 feet high, often forming clumps; stems smooth; flat, linear leaves (up to 48 inches long and to 1½ inches wide) with parallel veins on surfaces, prominent midrib, smooth surfaces, very rough margins, leaf sheaths also smooth with rough margins; open terminal inflorescence (panicle, up to 24 inches long and to 6 inches wide) with nodding branches, bearing female spikelets at ends and male spikelets near base of same branches; yellowish seed.

Flowering period: May through July.

Habitat: Slightly brackish and tidal fresh marshes; nontidal marshes.

Wetland indicator status: OBL.

Range: Maryland and Kentucky south to Florida, west to Texas and Oklahoma.

SEE ALSO salt and brackish marsh grasses, because most also extend into tidal fresh marshes.

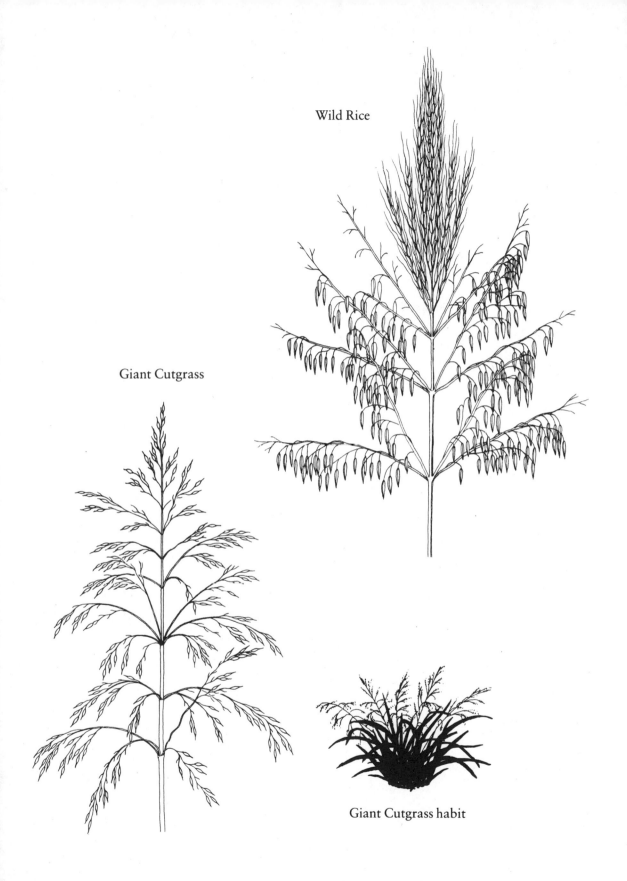

Wild Rice

Giant Cutgrass

Giant Cutgrass habit

SEDGES

Tussock Sedge

Carex stricta Lam.

Sedge Family
Cyperaceae

Description: Medium-height, erect, perennial grasslike plant, 1½–3½ feet tall, forming large clumps called tussocks; stems slender and three-angled; elongate, stiff, linear leaves (up to 2½ feet long, ¼ inch wide) with rough margins, tapering to a tip, channeled and rough above and keeled below, leaf sheath closed at back, leaves arising from base of tussock; inconspicuous flowers borne in two types of spikes—male spikes terminal, one to three in number, and female spikes axillary, two to six in number, covered by overlapping reddish brown or purplish brown scales; nutlet (achene) enclosed by an inflated sac (perigynium).

Flowering period: May through August.

Habitat: Tidal fresh marshes; nontidal marshes, swamps, and wet swales.

Wetland indicator status: OBL.

Range: Quebec and Nova Scotia to Minnesota, south to North Carolina, Tennessee, and Iowa.

Similar species: Shoreline Sedge (*C. hyalinolepis*) is not tussock-forming but has elongate, cylinder-shaped spikes that are somewhat similar in general appearance; it has septate leaves (visible on lower surface); it is mostly a Coastal Plain species occurring from southern New Jersey to northern Florida to Texas; it is OBL.

Bristlebract Sedge

Carex tribuloides Wahlenb.

Sedge Family
Cyperaceae

Description: Medium-height, perennial grasslike plant, up to 4 feet tall; stout, triangular stems, rough above, occurring in dense clumps; long, narrow-pointed, soft, linear leaves (up to 8 inches long and ⅛–¼ inch wide) with veined sheaths; terminal inflorescence of five to fifteen egg-shaped, bristly spikes with narrow perigynia (about three times as long as wide).

Flowering period: June into September.

Habitat: Tidal fresh marshes; nontidal marshes, wet meadows, and forested wetlands.

Wetland indicator status: FACW+.

Range: New Brunswick to Saskatchewan, south to Florida and Oklahoma.

Similar species: Pointed Broom Sedge (*C. scoparia*) is quite similar but has narrower leaves (⅛ inch or less); it is FACW. Two other sedges with egg-shaped spikes have wider, somewhat oval-shaped perigynia (about 1¾ times as long as wide, or less): Greenish-white Sedge (*C. albolutescens*; FAC+) and *C. alata*; they may be more common in southern tidal fresh marshes and swamps than the other two. Consult a taxonomic reference to verify your determination.

Tussock Sedge

Tussock Sedge habit

Bristlebract Sedge

Fox Sedge

Carex vulpinoidea Michx.

Sedge Family
Cyperaceae

Description: Medium-height, perennial grasslike plant, up to 3½ feet tall; stiff, triangular stems, very rough above, occurring in clumps; flat, rough, linear leaves usually exceed the stems; terminal flowering inflorescence (up to 6 inches long) composed of many cylindrical spikes, each subtended by a bristle (up to 2 inches long).

Flowering period: May through July.

Habitat: Tidal fresh marshes; nontidal marshes, wet meadows, and other wet places.

Wetland indicator status: OBL.

Range: Newfoundland to Pacific coast, south to Florida and California.

Similar species: Resembles Stalk-grain Sedge (*C. stipata*), which has somewhat winged triangular stems, and the upper part of its leaf sheath is usually wrinkled; it is OBL.

Fox Sedge

Saw Grass

Cladium jamaicense Crantz

Sedge Family
Cyperaceae

Description: Tall, coarse, perennial grasslike plant, up to 10 feet high, growing in dense clumps; stolons large (about ½ inch wide) and covered by overlapping scalelike leaves; stems somewhat triangular; long, tapering leaves (up to 3½ feet long and about ½ inch wide) with short, saw-toothed margins and narrow, fine-pointed tips; inconspicuous flowers borne in reddish brown spikelets in clusters of two to six at ends of often drooping branches forming a terminal inflorescence (to 32 inches long); olive or purplish, somewhat round nutlets with narrow tips.

Flowering period: July through October.

Habitat: Slightly brackish and tidal fresh marshes; nontidal marshes and hydric hammocks. (*Note:* This is the characteristic plant of the Florida Everglades.)

Wetland indicator status: OBL.

Range: Southeastern Virginia to Florida, west to Texas; also in the West Indies.

Similar species: Twig Rush (*C. mariscoides*) may occur in brackish and tidal fresh marshes; it grows to about 3 feet tall, has narrow grasslike leaves (up to 8 inches long and to about ⅛ inch wide) with weakly rough margins, and bears clusters of brown spikelets on upright branches; it is OBL.

Sheathed Flatsedge

Cyperus haspan L.

Sedge Family
Cyperaceae

Description: Annual grasslike plant, up to 2½ feet tall, with reddish roots; stems triangular and soft; linear basal or lower stem leaves (about ¼ inch wide) usually absent, two or three leaflike bracts at top of stem; many inconspicuous flowers, in flattened spikelets borne in umbellike clusters on long stalks (up to 6 inches long) forming a terminal inflorescence, overlapping flower scales red- or purple-tinged with rough margins and sharp-pointed tips; three-sided or lens-shaped nutlets covered with minute bumps.

Flowering period: July through September.

Habitat: Brackish and tidal fresh marshes; nontidal marshes, ditches, swales, wet to moist fields, shallow water, and edges of swamps.

Wetland indicator status: OBL.

Range: Southeastern Virginia to Florida, west to Texas; also in Tennessee and tropical America.

Similar species: Jointed Flatsedge (*C. articulatus*) has round to roundish triangular, distinctly jointed stems; it is OBL. Some other flatsedges of coastal marshes include Green Flatsedge (*C. virens*), Flat-leaf Flatsedge (*C. planifolius*; in Florida only), LeConte's Flatsedge (*C. lecontei*), Swamp Flatsedge (*C. ligularis*), and Poorland Flatsedge (*C. compressus*). Consult technical references to identify these species.

Sheathed Flatsedge

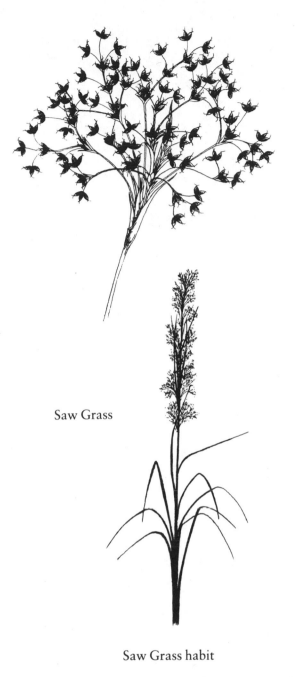

Saw Grass

Saw Grass habit

Twig Rush

Fragrant Galingale or Flatsedge

Cyperus odoratus L.

Sedge Family
Cyperaceae

Description: Medium-height, annual grasslike plant, up to 2 feet tall; stems triangular in cross-section; flat, linear basal and lower stem leaves (less than ½ inch wide) with spongy, purplish leaf sheaths and numerous leaflike bracts at top of stem subtending inflorescence; many inconspicuous flowers in flattened to somewhat rounded, yellow-brown spikelets borne in cylinder-shaped clusters on long stalks (up to 4 inches long) forming a terminal inflorescence, spikelet scales (⅛ inch long or less) brownish red with green midrib and rough margins; silver-brown to blackish three-sided nutlet (achene).

Flowering period: July through September.

Habitat: Tidal fresh marshes and edges of mangroves swamps; nontidal marshes, ditches, wet open areas, and exposed shores.

Wetland indicator status: FACW.

Range: Massachusetts to Minnesota, south to Florida and Texas; also in California and tropical America.

Similar species: See Umbrella Sedge (*C. strigosus*).

Retrorse Flatsedge

Cyperus retrorsus Chapm.

Sedge Family
Cyperaceae

Description: Perennial grasslike plant, up to 3½ feet tall; stems triangular with somewhat bulblike bases; rhizomes short and stout; linear basal and lower stem leaves (less than ½ inch wide), leaf sheaths sometimes red-spotted, usually three to seven leaflike bracts at top of stem subtending inflorescence; inconspicuous flowers in dull greenish spikelets borne in dense cylinder-shaped spikes (1 inch long or less) on stalks (up to 6 inches long), spikelets wide-spreading and lower ones downward-pointing (retrorse), flower scales (⅕ inch or less) greenish yellow or reddish with green ridges; reddish or olive-colored three-sided nutlets.

Flowering period: July into October.

Habitat: Sandy brackish and tidal fresh marshes; nontidal wetlands, dry sandy uplands, dunes, and disturbed sites.

Wetland indicator status: FACU+.

Range: Southeastern New York south to Florida, west to Texas and Oklahoma; also in Kentucky, Tennessee, and Missouri.

Similar species: Four-angled Flatsedge (*C. tetragonus*) has open cylindrical spikes, and its flower scales are more than ⅕ inch long; it is FAC+.

Straw-colored Umbrella Sedge

Cyperus strigosus L.

Sedge Family
Cyperaceae

Description: Low to medium-height, erect, perennial grasslike plant, 8–40 inches tall; rhizomes bulblike; stems thick, solid, smooth, and triangular in cross-section; elongate linear leaves arranged in three ranks, uppermost leaves (actually bracts) arranged in a cluster at top of stem and immediately below flowering inflorescence; inconspicuous scale-covered flowers borne in cylinder-shaped spikes (½–1½ inches long) forming a terminal inflorescence (umbel), spikes much branched with numerous horizontally radiating or erect, flattened, yellowish spikelets arranged along slender stalks; three-sided nutlet (achene).

Flowering period: July into October.

Habitat: Tidal fresh marshes; moist fields, swales, nontidal marshes, swamps, and wet shores.

Wetland indicator status: FACW.

Range: Quebec and Maine to Minnesota and South Dakota, south to Florida and Texas; also on Pacific coast, Washington to California.

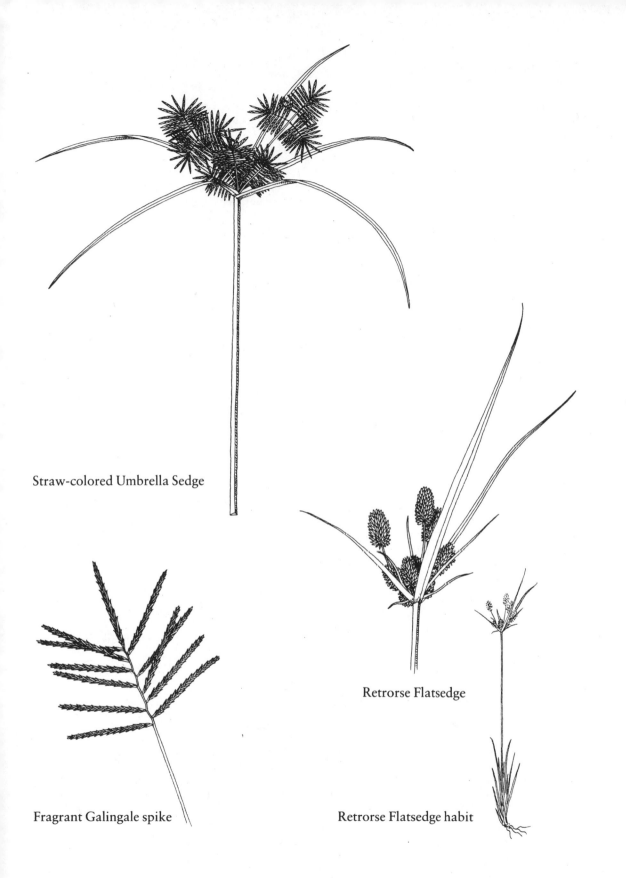

Straw-colored Umbrella Sedge

Fragrant Galingale spike

Retrorse Flatsedge

Retrorse Flatsedge habit

White-top Sedge

Dichromena colorata (L.) Hitchc.

Sedge Family
Cyperaceae

Description: Low to medium-height, perennial grasslike plant, up to 2 feet tall; rhizomes elongate and slender; stems triangular; mostly basal leaves with smooth surfaces and rough margins, long, tapering, flat stem leaves (to 16 inches long) usually occurring below middle of stem; flowerlike terminal inflorescence (about ½ inch wide) composed of clusters of ten to twenty whitish headlike spikelets subtended by downward-pointing leaflike bracts, whitish at base with greenish tips; lens-shaped nutlet with pitted surfaces and truncate-triangular tubercle (beak).

Flowering period: May through September.

Habitat: Slightly brackish and tidal fresh marshes and upper edges of salt marshes; sand flats, nontidal marshes, savannahs, pine flatwoods, bogs, and ditches.

Wetland indicator status: FACW.

Range: Virginia to Florida, west to Texas; also in Mexico and the West Indies.

Similar species: Giant White-top Sedge (*D. latifolia*), a nontidal relative, usually has seven to ten leaflike bracts in the inflorescence; it is FACW+.

Three-way Sedge

Dulichium arundinaceum (L.) Britton

Sedge Family
Cyperaceae

Description: Low to medium-height, erect, perennial grasslike plant, up to 3½ feet tall, commonly less than 2 feet; stem hollow, round to roundish triangular, and jointed; numerous linear leaves (2–5 inches long and less than ⅓ inch wide) distinctly three-ranked, lower leaves bladeless, upper leaf sheaths frequently overlapping; inconspicuous flowers covered by brownish scales, borne on several spikes (less than 1¼ inches long) with short stalks (peduncles, less than 1 inch long) from upper leaf axils; flattened nutlet (achene) with six to nine barbed bristles.

Flowering period: July to October.

Habitat: Irregularly flooded tidal fresh marshes; nontidal marshes, bogs, swamps, and margins of ponds.

Wetland indicator status: OBL.

Range: Newfoundland to Minnesota, south to Florida and Texas; also Montana to Washington.

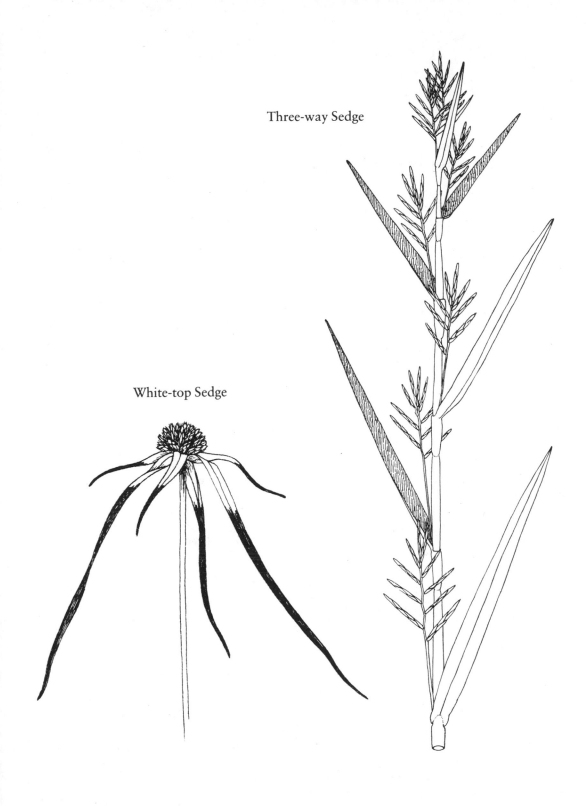

Three-way Sedge

White-top Sedge

Long-tubercle Spike-rush

Eleocharis tuberculosa (Michx.) Roem.
  & J. A. Schultes

Sedge Family
Cyperaceae

Description: Annual or perennial, medium-height, grasslike plant, up to 2 feet tall; stem roundish in cross-section and ribbed, apparently leafless, leaves reduced to basal sheaths; single budlike spikelet (⅕–⅗ inch long and much wider than stem) located at top of stem, spikelet covered by brownish scales with green midrib; olive-colored, three-sided nutlets (achenes) with whitish, spongy, somewhat pyramid-shaped tubercle (about equal in thickness and length to the achene).

Flowering period: June through September.

Habitat: Brackish and tidal fresh marshes; nontidal marshes, bogs, savannahs, ditches, and borders of swamps.

Wetland indicator status: FACW+.

Range: Massachusetts south to Florida, west to Texas.

Similar species: See Gulf Coast Spike-rush (*E. cellulosa*). Horsetail Spike-rush (*E. equisetoides*) also has roundish stem in cross-section but with conspicuous cell walls (septate); its spikelet is about the same width as its stem; it is OBL. Other spike-rushes of tidal marshes with three-sided nutlets are Dwarf Spike-rush (*E. parvula*), which is less than 6 inches tall; Creeping Spike-rush (*E. fallax*) with somewhat rounded stems, reddish upper leaf sheaths, and reddish rhizomes; and Beaked Spike-rush (*E. rostellata*) with flattened stems in thick clumps. All three are OBL, and spike-rushes, in general, range from FACW to OBL in wetland indicator status. Other spike-rushes have lens-shaped nutlets. Iris-leaf Yellow-eyed Grass (*Xyris iridifolia*) has a budlike terminal spikelet with conspicuous but small yellow-petaled flowers and narrow linear basal leaves; it is OBL. Southern Umbrella-grass (*Fuirena scirpoidea*) also has a budlike terminal spikelet but has several leaf sheaths (actually reduced bladeless leaves) arranged along its stem; it is OBL.

Tall Beak-rush

Rhynchospora macrostachya Torr.

Sedge Family
Cyperaceae

Description: Perennial grasslike plant, up to 5 feet high, occurring in clumps; stems triangular; linear leaves (⅕–⅗ inch wide) usually with rough margins and smooth surfaces; inconspicuous flowers borne in usually ten or more reddish or brownish spikelets (less than 1 inch long) in dense clusters forming an open inflorescence with numerous short branches; flat brownish nutlet with elongate, minutely barbed tubercle (beak) and long bristles (longer than nutlet).

Flowering period: July into October.

Habitat: Slightly brackish and tidal fresh marshes; nontidal marshes, muddy shores, margins of lakes, ponds, and rivers, and ditches.

Wetland indicator status: OBL.

Range: Eastern Massachusetts south to northern Florida, west to eastern Texas, and north to southern Michigan.

Similar species: Short-bristle Beak-rush (*R. corniculata*) usually has only two to seven spikelets per cluster and its nutlet has a long tubercle, but its bristles are usually much shorter or as long as the nutlet; it is OBL. Fasciculate Beak-rush (*R. fascicularis*) has a nutlet lacking a long tubercle but having a triangle-shaped beak; it is FACW+.

Long-tubercle Spike-rush   Tall Beak-rush spikelet clusters

Wool Grass

Scirpus cyperinus (L.) Kunth

Sedge Family
Cyperaceae

Description: Medium-height to tall, erect, perennial grasslike plant, up to 6½ feet high, commonly 4–5 feet, growing in dense clumps; stem roundish to weakly triangular, especially near base; simple, elongate, rough-margined linear leaves (less than ½ inch wide), drooping at tips, dense cluster of basal leaves present; numerous inconspicuous flowers covered by reddish brown scales borne on mostly sessile budlike spikelets (usually ⅕ inch long) covered by cottonlike hairs (at maturity) and clustered (three to fifteen) in terminal inflorescence, somewhat drooping at maturity, and subtended by spreading and drooping leafy bracts; yellow-gray to white nutlet (achene).

Flowering period: July through September.

Habitat: Irregularly flooded tidal fresh marshes; nontidal marshes, wet meadows, and swamps.

Wetland indicator status: OBL.

Range: Newfoundland to Minnesota, south to Florida and Louisana.

Common Three-square

Scirpus pungens Vahl
[*Scirpus americanus* Pers.]

Sedge Family
Cyperaceae

Description: Medium-height, erect perennial herb, up to 4 feet tall; rhizomes hard and elongate; stems stout and triangular in cross-section, occasionally twisted; no apparent leaves but actually one to three stemlike erect leaves (up to 16 inches long); inconspicuous flowers borne in several, often three or four, sessile budlike spikelets covered by brown scales located near top of stem (portion above spikelets is 1¼–5 inches long); gray to black nutlet (achene).

Flowering period: June into September.

Habitat: Brackish and tidal fresh marshes (regularly and irregularly flooded zones) and upper borders of salt marshes where freshwater influence is strong; wet sandy shores, nontidal marshes, and shallow waters.

Wetland indicator status: OBL.

Range: Newfoundland, Quebec, and Minnesota south to Florida and Texas; also in western states to the Pacific coast.

Similar species: Olney's Three-square (*S. americanus*, formerly *S. olneyi*) has triangular stems with deeply concave sides; it occurs in brackish marshes; it is OBL.

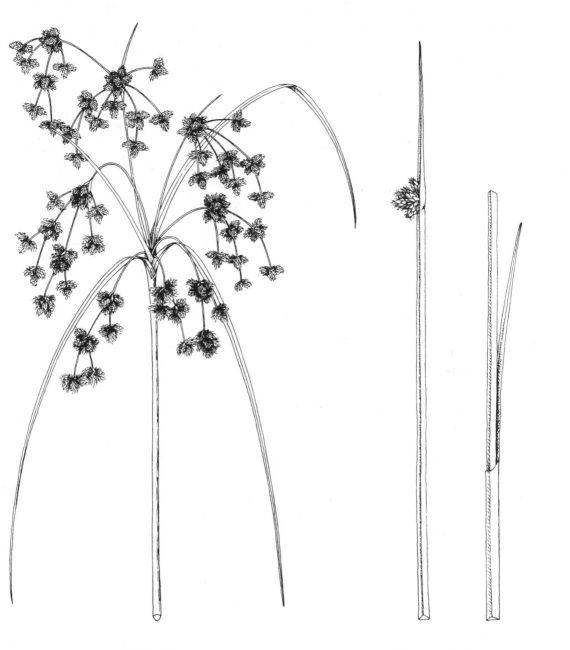

Wool Grass   Common Three-square

Soft-stemmed Bulrush

Scirpus validus Vahl

Sedge Family
Cyperaceae

Description: Tall, erect perennial herb, up to 10 feet high, forming dense colonies; rhizomes slender; stems soft, round in cross-section, tapering to a point, usually grayish green; no apparent leaves, leaves reduced to basal leaf sheaths; inconspicuous flowers borne in an open inflorescence of many stalked budlike spikelets (⅕–⅘ inch long) covered by reddish brown scales located immediately below top of stem, spikelet clusters mostly drooping, few erect; brownish gray nutlet (achene).

Flowering period: June into September.

Habitat: Brackish and tidal fresh marshes (regularly and irregularly flooded zones); inland shallow waters, muddy shores, and nontidal marshes.

Wetland indicator status: OBL.

Range: Newfoundland to Florida, west to the Pacific coast.

Similar species: Hard-stemmed Bulrush (*S. acutus*) has dark green, hard, round stems, stout rhizomes, and stalked spikelets forming a nearly terminal inflorescence; it is OBL. California Bulrush or Bullwhip (*S. californicus*) has bluntly triangular stems with a few basal leaves reduced to brownish sheaths with fringed margins; it is OBL. Canby's Bulrush (*S. etuberculatus*) has drooping spikelet clusters at the top of its stem, but its stem is three-angled and leafy; it is OBL.

Soft-stemmed Bulrush

RUSHES

Canada Rush

Juncus canadensis J. Gay

Rush Family
Juncaceae

Description: Medium-height, erect, perennial grasslike plant, 16–40 inches tall, often growing in small clumps; elongate linear leaves, round in cross-section with partitions (transverse septa) at regular intervals; five to fifty inconspicuous flowers borne in somewhat rounded heads (glomerules) on compact to open, erect terminal and axillary inflorescences (up to 8 inches long); fruit capsule bearing many minute seeds.

Flowering period: July into October.

Habitat: Slightly brackish and tidal fresh marshes, occasionally borders of salt marshes; nontidal marshes, swamps, and wet shores.

Wetland indicator status: OBL.

Range: Quebec to Nova Scotia to Minnesota, south to Georgia and northern Illinois.

Similar species: Closely resembling *J. canadensis* are White-root Rush (*J. brachycarpus*; FACW), Many-head Rush (*J. polycephalus*; OBL), Round-head Rush (*J. validus*; FACW+), and Tapertip Rush (*J. acuminatus*; OBL). See Needle-pod Rush (*J. scirpoides*). Slim-pod Rush (*J. diffusissimus*; FACW) and Red-pod Rush (*J. trigonocarpus*; OBL) have loose, turban-shaped (half-rounded) clusters borne on many-branched open inflorescences. Rushes with nonseptate leaves found in coastal marshes are Grass-leaf Rush (*J. marginatus*; FACW) with somewhat flattened stems and Toad Rush (*J. bufonius*; FACW), which is less than 1½ feet tall and has an inflorescence comprising more than half its height. See Soft Rush (*J. effusus*). Identification of rushes requires examination of seeds and fruit capsules, so refer to a taxonomic manual for specifics.

Canada Rush

Soft Rush

Juncus effusus L.

Rush Family

Juncaceae

Description: Medium-height, erect, perennial grasslike herb, up to 3½ feet tall, forming dense clumps or tussocks; stems stout, unbranched, round in cross-section, soft, with regular vertical fine lines (ribs), sheathed (usually brown) at base, up to 8 inches long with a bristle tip; no apparent leaves, leaves actually basal sheaths; inconspicuous greenish brown scaly flowers borne in somewhat erect clusters arising from a single point on the upper half of the stem (lateral inflorescence); fruit capsule containing many minute seeds. (*Note:* Much of stems remain greenish through winter.)

Flowering period: June into September.

Habitat: Tidal fresh marshes; nontidal marshes, wet meadows, shrub swamps, and wet pastures.

Wetland indicator status: FACW+.

Range: Newfoundland to North Dakota, south to Florida and Texas.

Similar species: Leathery Rush (*J. coriaceous*) has somewhat rounded (in cross-section) leaves, flexible stems, and loose, few-branched lateral inflorescences; it is FACW.

Needle-pod Rush

Juncus scirpoides Lam.

Rush Family

Juncaceae

Description: Low to medium-height, perennial grasslike plant up to 3 feet tall; rhizomes; round, linear stems; two round (in cross-section) leaves (up to 4 inches long) with distinct internal chambers (septate, cell walls evident when crushed but visible externally); compact or branched terminal inflorescence (up to 6 inches long) bearing globe-shaped flowering heads; blunt seeds without long tails.

Flowering period: June into October.

Habitat: Sandy tidal marshes and shores; nontidal marshes, wet prairies, savannahs, and pond shores.

Wetland indicator status: FACW+.

Range: New York to Missouri, south to Florida and Texas.

Similar species: Big-head Rush (*J. megacephalus*) lacks rhizomes, the cell walls of its septate leaves are not seen externally, and its uppermost stem leaf is much reduced in size; sometimes only the sheath remains; it is OBL. See Canada Rush (*J. canadensis*).

Soft Rush   Needle-pod Rush

OTHER GRASSLIKE PLANTS

Sweet Flag
Acorus calamus L.

Arum Family
Araceae

Description: Medium-height, erect perennial herb, usually 1–4 feet tall, rarely to 7 feet; simple, entire, aromatic, linear, sword-shaped basal leaves with midrib slightly off-center; small yellow-brown flowers borne on an erect fleshy appendage (spadix, 2–4 inches long) developing from a leaflike peduncle (scape).

Flowering period: May to August.

Habitat: Tidal fresh marshes; shallow waters, nontidal swamps, and wet meadows.

Wetland indicator status: OBL.

Range: Nova Scotia and Quebec to Montana, Oregon, and Alberta, south to Florida, Texas, and Colorado; native of Eurasia.

Similar species: Leaves of *Iris* spp. may be confused with *Acorus*, but those leaves are not aromatic.

Eastern or Lesser Bur-reed
Sparganium americanum Nutt.

Bur-reed Family
Sparganiaceae

Description: Medium-height perennial, up to 3½ feet tall; simple, entire, soft, flat, linear leaves clasping stem to form sheaths at base, undersides somewhat triangular in cross-section, alternately arranged; minute greenish to whitish flowers borne in ball-shaped heads, arranged along a branched inflorescence; ball-like fruit clusters of nutlets.

Flowering period: May through August.

Habitat: Muddy shores, shallow waters and nontidal marshes; possibly tidal fresh marshes and shores.

Wetland indicator status: OBL.

Range: Newfoundland and Quebec to Minnesota, south to Florida and Louisiana.

Broad-leaved Cattail
Typha latifolia L.

Cattail Family
Typhaceae

Description: Medium-height to tall perennial herb, up to 10 feet high; pith white at base of stem; simple, entire, elongate linear basal leaves (up to 1 inch wide) sheathing at base and ascending along stem in an alternately arranged fashion; inconspicuous flowers borne on terminal spike composed of two parts, male flower spike above and contiguous with female spike.

Flowering period: March into July.

Habitat: Tidal fresh marshes; nontidal marshes, ponds, and ditches.

Wetland indicator status: OBL.

Range: Newfoundland to Alaska, south to Florida and Mexico.

Similar species: Narrow-leaved Cattail (*T. angustifolia*) has narrower, dark green leaves (to ½ inch wide and usually less than ten in number), a space between its male and female spike. Southern Cattail (*T. domingensis*) resembles Narrow-leaved Cattail, but it is taller (8–13 feet) and has more than ten yellowish green leaves. Blue Cattail (*T. glauca*) appears to be intermediate between *T. latifolia* and *T. angustifolia*; its pith is yellowish brown at the base of the stem. All cattails are OBL.

SEE ALSO Mudwort (*Limosella subulata*), spike-rushes (*Eleocharis* spp.), and Riverbank Quillwort (*Isoetes riparia*).

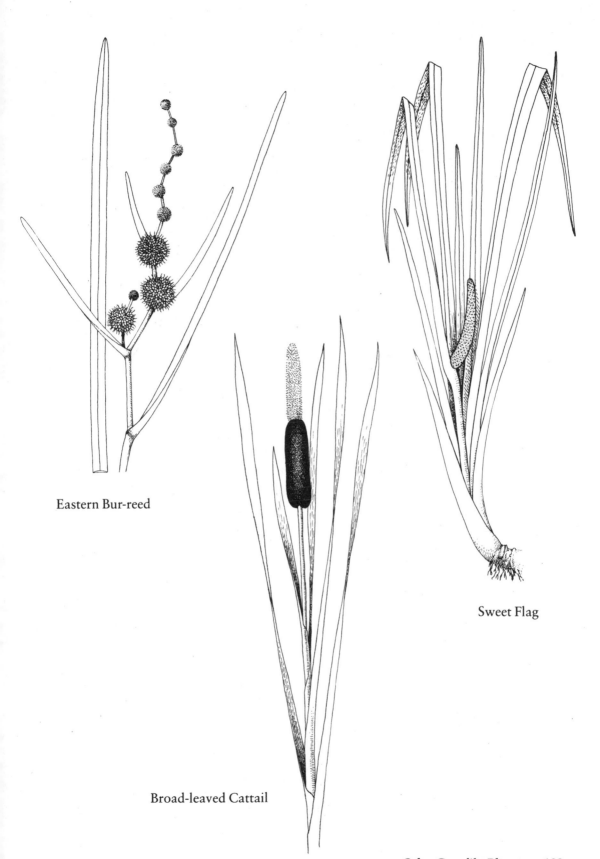

Eastern Bur-reed

Broad-leaved Cattail

Sweet Flag

FLESHY HERBS

Bull-tongue
or Lance-leaved Arrowhead
Sagittaria lancifolia L.
Water Plantain Family
Alismataceae

Description: Medium-height, somewhat fleshy-leaved perennial herb, up to 4½ feet tall, usually occurring in clumps and forming dense stands; leaves basal, young leaves narrow, nearly round in cross-section, older leaves erect, thickened, somewhat leathery, lance-shaped (up to 16 inches long and to 4 inches wide), borne on long spongy stalks (to more than 8 inches long); many white three-petaled flowers (about 1½ inches wide) borne on stalks (up to 1 inch long) in clusters of three arranged in up to twelve whorls on a separate flowering stem (scape); fruiting heads (about ¾ inch wide) bearing somewhat sickle-shaped nutlets.

Flowering period: May into October.

Habitat: Slightly brackish and tidal fresh marshes; nontidal marshes, muddy shores, and swamps.

Wetland indicator status: OBL.

Range: Delaware and Maryland south along the Coastal Plain to Florida, west to Texas and Oklahoma; also in tropical America.

Similar species: Coastal Arrowhead (*S. falcata*) is essentially the same and has been combined with *S. lancifolia* by some authors; its bracts and sepals subtending the flowers are covered with minute bumps (papillose). Delta Arrowhead (*S. platyphylla*) occurs in similar habitats along the Gulf coast; it has longer, wider, somewhat egg-shaped leaves (to 7¼ inches long and to 3¼ inches wide) that overtop the flowering stem, and its lower flower stalks are nodding. All arrowheads are OBL.

Golden Club
Orontium aquaticum L.
Arum Family
Araceae

Description: Medium-height, erect, fleshy perennial herb, up to 1½ feet tall; rhizomes stout, fleshy; simple, entire, egg-shaped, fleshy basal leaves (3–10 inches long and about a third as wide) tapering distally to a pointed tip, toward base rolled inwardly where attached to long, fleshy stalk (petiole, up to 8 inches long); numerous minute yellow flowers borne at end of separate fertile, showy, fleshy stalk (spadix), whitish below and surrounded by a tubular leaf at base.

Flowering period: March into May.

Habitat: Muddy shores of regularly flooded tidal fresh marshes; shallow waters and inland shores.

Wetland indicator status: OBL.

Range: Massachusetts and central New York south to Florida and Kentucky.

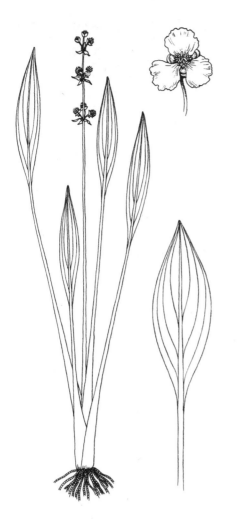

Bull-tongue Arrowhead

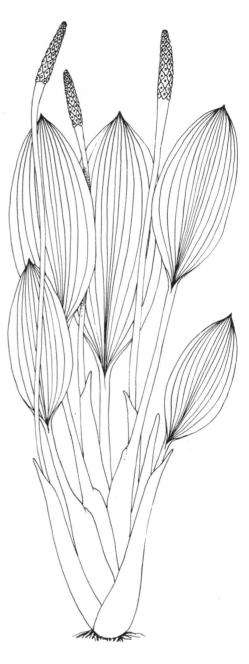

Golden Club

Arrow Arum or Tuckahoe

Peltandra virginica (L.) Kunth

Arum Family
Araceae

Description: Low to medium-height, erect, fleshy perennial herb, up to 2 feet tall; simple, entire, triangular-shaped, thick, fleshy basal leaves (4–12 inches long at flowering and growing larger afterward), ends of basal lobes rounded or pointed, three-nerved, on long petioles; inconspicuous yellowish flowers borne on a fleshy spike (spadix) enclosed within a pointed, leaflike, green fleshy structure (spathe); greenish, slimy, and pealike berry.

Flowering period: May to July.

Habitat: Tidal fresh marshes and swamps and slightly brackish marshes (regularly and irregularly flooded zones); nontidal swamps, marshes, and shallow waters of ponds and lakes.

Wetland indicator status: OBL.

Range: Southern Maine and southwestern Quebec to Michigan, southern Ontario, and Missouri, south to Florida and Texas.

Butterweed

Senecio glabellus Poir.

Composite or Aster Family
Compositae (*Asteraceae*)

Description: Erect, fleshy-leaved annual herb, up to 3¼ feet tall; stems soft, hollow, and smooth; compound, round- to sharp-toothed, somewhat lance-shaped to egg-shaped, fleshy leaves (up to 8 inches long and to 3 inches wide), lower leaves stalked, upper leaves stalkless, basal, and alternately arranged; small, showy, yellow daisylike flowers borne in clusters; brown nutlets (about ⅗ inch long).

Flowering period: March into June.

Habitat: Tidal fresh marshes; nontidal forested wetlands, wet meadows, hydric hammocks, floodplain forests, and ditches.

Wetland indicator status: FACW+.

Range: Southeastern North Carolina to Florida, west to eastern Texas, north to southern Illinois, Ohio, and South Dakota.

Pygmyweed

Crassula aquatica (L.) Schoenl.
[*Tillaea aquatica* L.]

Orpine Family
Crassulaceae

Description: Low-growing, erect, annual fleshy herb, up to 4 inches tall; stem branched from base; simple, linear, fleshy sessile leaves (usually ⅕ inch long), oppositely arranged and joined at stem; minute white or greenish white four-petaled flowers borne singly in leaf axils.

Flowering period: May through August.

Habitat: Regularly flooded mud flats along brackish and tidal fresh marshes; mud flats along pools and shores.

Wetland indicator status: OBL.

Range: Quebec and Newfoundland to Maryland; also from Louisiana to Texas and along the Pacific coast; also in Mexico.

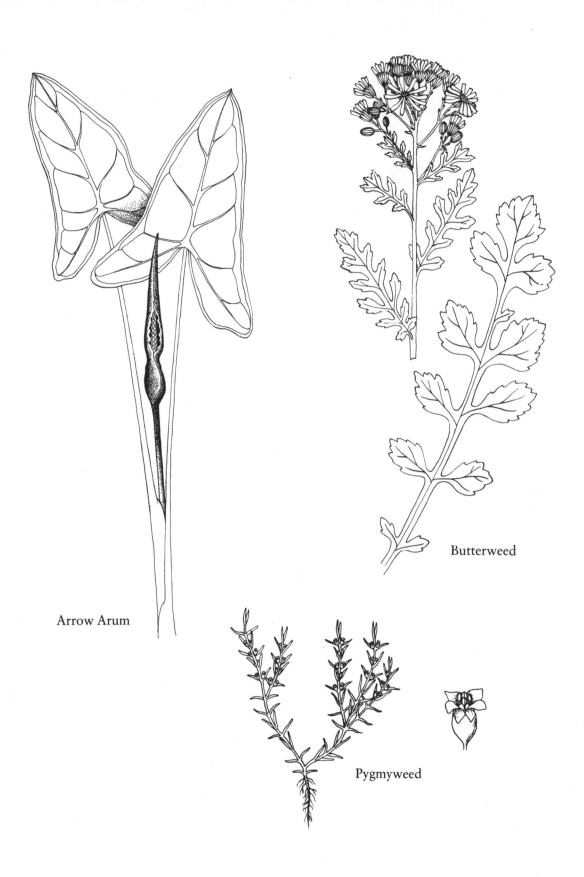

Arrow Arum

Butterweed

Pygmyweed

Riverbank Quillwort

Isoetes riparia Engelm. ex A. Braun
[includes *Isoetes saccharata* Engelm.]

Quillwort Family
Isoetaceae

Description: Low-growing, somewhat fleshy-leaved, erect perennial herb, 3½–12 inches tall; stem appearing absent but actually reduced to fleshy bulblike corm; numerous elongate, hollow, erect, linear leaves (up to 12 inches long), sharp-pointed, usually pale green, divided into four air cavities (in cross-section) and separated along leaf length by horizontal cell walls, leaf bases greatly swollen; sporangia borne at base of leaves.

Fruiting period: May to October.

Habitat: Mud flats or gravelly shores along regularly flooded tidal freshwater wetlands; inland shores.

Wetland indicator status: OBL.

Range: Maine to South Carolina.

Similar species: To distinguish from other quillworts requires examination of spores. *I. riparia* is, however, the common tidal species.

Spatterdock

Nuphar luteum (L.) Sibth. & J. E. Smith
[*Nuphar advena* Ait.]

Water Lily Family
Nymphaeaceae

Description: Low to medium-height, erect, perennial, fleshy herb, up to 16 inches tall; simple, entire, heart-shaped, fleshy basal leaves (up to 20 inches long and wide), basal lobes separated by a broadly triangular sinus, borne on rounded stalks (petioles); single yellow flower (1½–2½ inches wide) with usually five or six "petals" borne on a long fleshy stalk (peduncle).

Flowering period: April to October.

Habitat: Tidal fresh marshes; nontidal marshes, swamps, and ponds.

Wetland indicator status: OBL.

Range: Southern Maine to Wisconsin and Nebraska, south to Florida and Texas.

Similar species: Pickerelweed (*Pontederia cordata*) occurs in the same habitats; it has numerous violet-blue flowers borne on a terminal stalk, and its leaf stalks (petioles) do not form a distinct midrib on the underside of the leaf as in *Nuphar luteum*; it is OBL. Narrow-leaved Yellow Pond Lily (*Nuphar luteum* spp. *sagittifolium*) has floating and submerged leaves, and its floating leaves are more than three times as long as wide; it is OBL. Elephant-ear (*Colocasia antiquorum*), an exotic emergent herb, has leaves resembling those of *Nuphar luteum*, but the former's are much larger (up to 3 feet long); it has been reported in Louisiana's tidal fresh marshes; it is OBL.

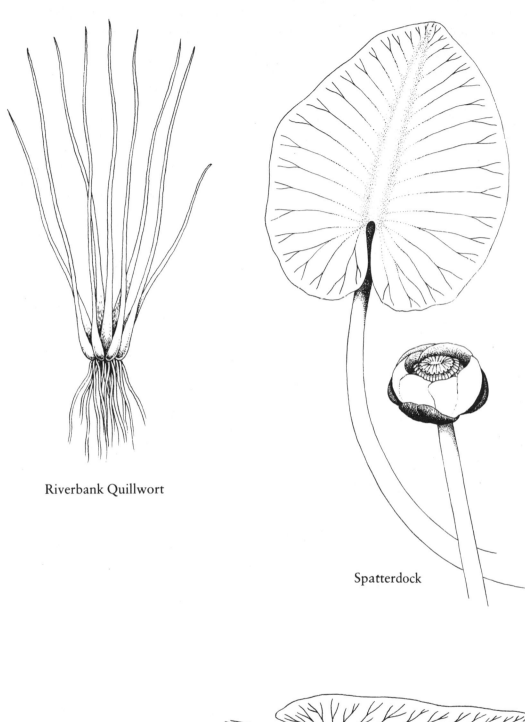

Riverbank Quillwort

Spatterdock

Narrow-leaved Yellow Pond Lily

Kidney-leaf Mud Plantain

Heteranthera reniformis Ruiz & Pavon

Pickerelweed Family
Pontederiaceae

Description: Short, creeping perennial herb or floating (sometimes submerged) aquatic plant; simple, entire, thickened, somewhat fleshy, kidney-shaped to heart-shaped basal leaves (up to 3 inches long and to 3 inches wide) with somewhat heart-shaped bases, long-stalked (up to 6 inches long); one to many small, bluish, six-"petaled," star-shaped tubular flowers (less than ½ inch long) surrounded by leafy bract (spathe, ½–1¼ inches long); three-valved oval fruit capsules (up to ⅘ inch long) bearing numerous ridged seeds.

Flowering period: June through October.

Habitat: Edges of tidal fresh marshes, intertidal mud flats, and shallow tidal waters; muddy shores and shallow inland waters.

Wetland indicator status: OBL.

Range: Connecticut and New York south to Florida, west to Texas and Nebraska; also in tropical America.

Similar species: Water Star-grass (*Zosterella dubia*, formerly *Heteranthera dubia*) has alternately arranged, linear, threadlike leaves and six-"petaled" yellow tubular flowers borne singly in a leafy spathe; it is OBL.

Pickerelweed

Pontederia cordata L.

Pickerelweed Family
Pontederiaceae

Description: Medium-height, erect, fleshy perennial herb, up to 3½ feet tall; simple, entire, thick, fleshy, heart-shaped, occasionally lance-shaped leaves (up to 7¼ inches long) on long petioles, basal and alternately arranged; numerous small violet-blue tubular flowers with three upper lobes (united) and three lower lobes (separated) borne on terminal spikelike inflorescence (3–4 inches long).

Flowering period: March to November.

Habitat: Tidal fresh marshes, occasionally slightly brackish marshes; nontidal marshes and shallow waters of ponds and lakes.

Wetland indicator status: OBL.

Range: Nova Scotia to Ontario and Minnesota, south to northern Florida and Texas.

Similar species: Spatterdock (*Nuphar luteum*) has heart-shaped leaves with a midrib underneath formed by a continuation of the petiole; it also bears a single yellow flower; it is OBL.

SEE ALSO Lizard's Tail (*Saururus cernuus*), Jewelweed (*Impatiens capensis*), Coastal Water-hyssop (*Bacopa monnieri*), Marsh Pennywort (*Hydrocotyle umbellata*), Elongated Lobelia (*Lobelia elongata*), White Boltonia (*Boltonia asteroides*), and Pink Ammania (*Ammania latifolia*).

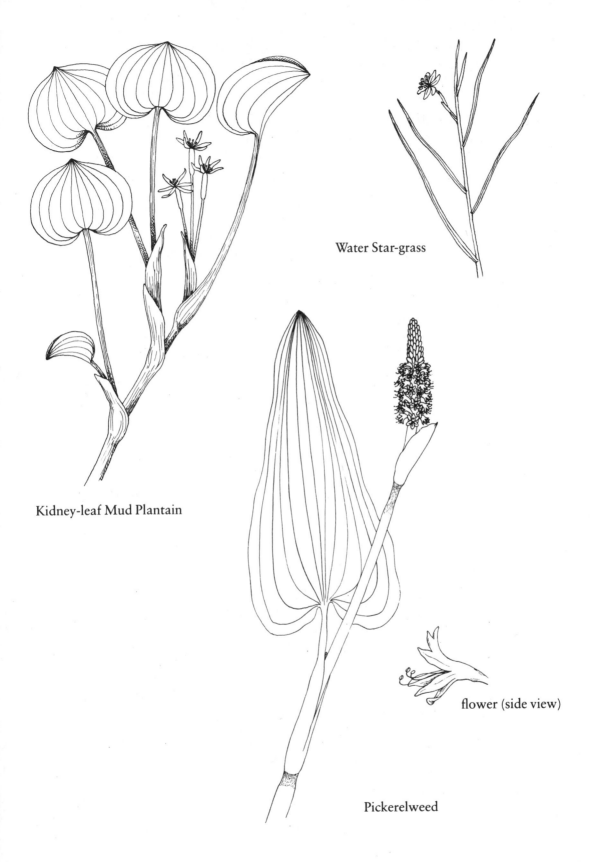

Water Star-grass

Kidney-leaf Mud Plantain

flower (side view)

Pickerelweed

FLOWERING HERBS WITH BASAL LEAVES ONLY

Southern Water Plantain or Mud Plantain

Alisma subcordatum Raf.

Water Plantain Family
Alismataceae

Description: Medium-height, erect perennial herb, up to about 2 feet tall; simple, entire, somewhat thickened, oval-shaped, mostly basal leaves (up to 8 inches long and to 6 inches wide) with abruptly pointed tips and somewhat heart-shaped bases, long-stalked and sheathing at base; small three-petaled white (sometimes pinkish) flowers (less than ⅓ inch wide) borne on separate-stalked fertile inflorescence with whorled branches; fruit cluster of flattened seeds.

Flowering period: April into November.

Habitat: Tidal fresh marshes and shores; non-tidal marshes, stream borders, ponds, and ditches.

Wetland indicator status: OBL.

Range: New England and Ontario to Minnesota, south to Florida and Texas.

Big-leaved Arrowhead or Wapato

Sagittaria latifolia Willd.

Water Plantain Family
Alismataceae

Description: Medium-height, erect perennial herb, up to 4 feet tall; simple, entire, basal leaves (2–16 inches long, 1–10 inches wide) broadly to narrowly arrowhead-shaped; white three-petaled flowers (1–1½ inches wide) arranged in whorls of two to fifteen, borne on single elongate stalk (peduncle, up to 4 feet tall); fruits green nutlets (achenes) joined in ball-shaped clusters.

Flowering period: June through September.

Habitat: Tidal fresh marshes; nontidal marshes and swamps, borders of streams, lakes, and ponds.

Wetland indicator status: OBL.

Range: Nova Scotia to British Columbia, south to Florida, California, and Mexico.

Similar species: See Bull-tongue or Lance-leaved Arrowhead (*S. lancifolia*).

Awl-leaf Arrowhead

Sagittaria subulata (L.) Buchenau

Water Plantain Family
Alismataceae

Description: Perennial flowering aquatic herb, submergent or low emergent, often forming extensive colonies in tidal waters; simple, entire, thick, linear basal leaves (usually less than 12 inches long and to about ½ inch wide), somewhat lens-shaped in cross-section, raised veins absent; small white three-petaled flowers (about ⅗ inch wide) borne on long stalks in one to ten whorls, stamen filaments smooth; nodding fruiting heads of green nutlets (achenes).

Flowering period: May through September.

Habitat: Brackish and tidal fresh waters, inter-tidal mud flats, and regularly flooded marshes.

Wetland indicator status: OBL.

Range: Massachusetts to Florida, west to Mississippi.

Similar species: Grass-leaved Arrowhead (*S. graminea*) has a similar form, but its male flowers have hairy filaments; it is OBL. Spring-tape Arrowhead (*S. kurziana*) has submerged ribbonlike leaves (up to 9 feet long) with one to five raised veins; it occurs in slightly brackish and tidal fresh waters in north-central Florida; it is OBL.

flower

Big-leaved Arrowhead

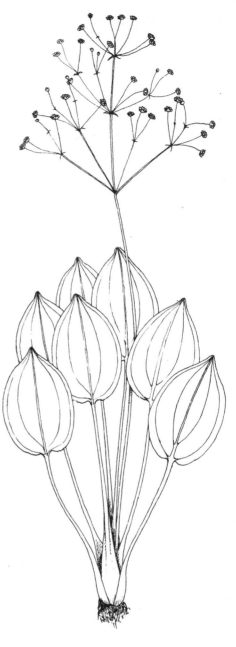

Southern Water Plantain

Awl-leaf Arrowhead

Southern Swamp Lily

Crinum americanum L.
Amaryllis Family
Amaryllidaceae

Description: Medium-height, perennial flowering herb, up to 5 feet tall, with bulbous bases; bulbs fleshy (up to 5 inches wide); simple, entire or soft-toothed, thick, elongate, internally chambered (septate), linear basal leaves (to 5 feet long and to 2 inches wide) overlapping at base in an alternate fashion; two to six large showy, fragrant, white (sometimes pinkish-tinged) six-"petaled" tubular flowers (less than 1 foot wide) borne in a terminal cluster (umbel), flower tube greenish and narrow (up to 5½ inches long); fruit capsules (about 1⅕ inches wide) thin-walled, roundish, with prominent beak bearing one or more fleshy seeds.

Flowering period: May through November.

Habitat: Tidal fresh marshes and swamps; nontidal marshes and swamps, shores of streams and lakes, and ditches.

Wetland indicator status: OBL.

Range: Georgia south along the Coastal Plain to Florida, west to Texas.

Similar species: See Marsh Spider Lily (*Hymenocallis crassifolia*).

Parker's Pipewort

Eriocaulon parkeri B. Rob.

Pipewort Family
Eriocaulaceae

Description: Low, erect perennial herb, 1–4 inches tall; thin, membranous, grasslike basal leaves (1–2½ inches long), linear, tapering to a fine tip; small white flowers in dense button-shaped head at end of four-angled stalk (peduncle or scape) that extends above leaves, usually two to four peduncles.

Flowering period: July to October.

Habitat: Tidal freshwater (occasionally slightly brackish) mud flats and shallow waters.

Wetland indicator status: OBL.

Range: Quebec and Maine to North Carolina.

Similar species: Two other pipeworts have been reported in southern tidal fresh waters: Ten-angle Pipewort (*E. decangulare*) and White Buttons (*E. septangulare*). The former species has a hard flower head, whereas the latter has a soft head that can be compressed upon squeezing; both are OBL. *E. decangulare* is the common pipewort along the southeastern Coastal Plain, where it grows to about 3½ feet tall.

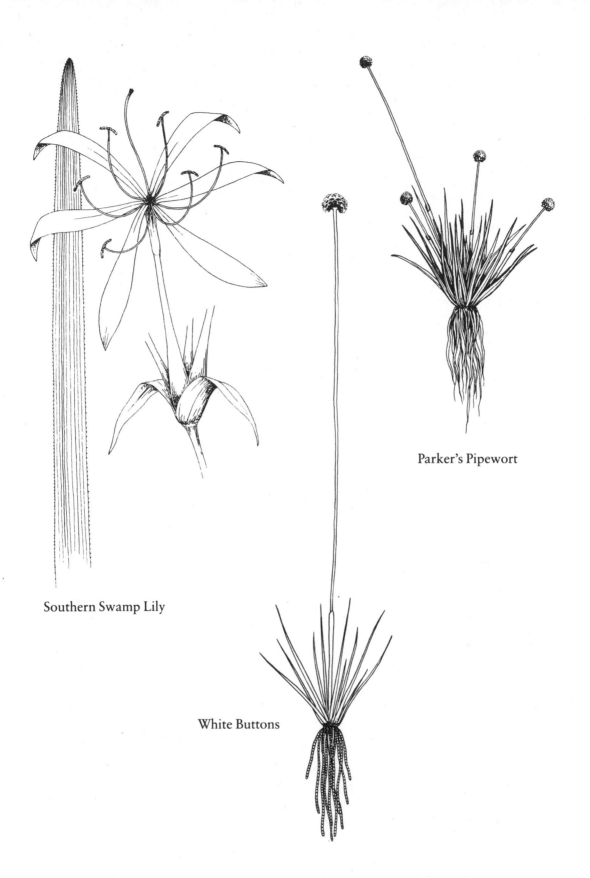

Southern Swamp Lily

Parker's Pipewort

White Buttons

Southern Blue Flag

Iris virginica L.

Iris Family
Iridaceae

Description: Medium-height, erect perennial herb, up to 3½ feet tall; rhizomes thick; stems usually one- or two-branched; simple, entire, green, sword-shaped basal leaves (up to 40 inches long and to 1⅕ inches wide) clasping at base; large showy, bluish purple to violet, six-"petaled" irislike flowers composed of three petals (2–3 inches long) surrounded by three larger, more colorful sepals (up to 4 inches long and to about 1½ inches wide), sepals marked with yellow near tips, flowers borne on separate stalk with leaflike bracts; somewhat angled three-parted fruit capsule (up to 3 inches long and to 1 inch wide).

Flowering period: April into June.

Habitat: Slightly brackish and tidal fresh marshes; nontidal marshes and swamps, wet savannahs and pinelands, wet meadows, ditches, and shallow water.

Wetland indicator status: OBL.

Range: Southeastern Virginia to Florida, west to Texas and Oklahoma.

Similar species: Northern Blue Flag (*I. versicolor*) is very similar, but has pale green leaves; it occurs from Virginia north; it is OBL. (*Note:* This is the illustrated species.) Yellow Flag (*I. pseudacorus*), an introduced species from Eurasia, has yellow flowers; it is OBL.

Mudwort

Limosella subulata E. Ives

Figwort Family
Scrophulariaceae

Description: Low, erect, annual grasslike herb, up to 2 inches tall; stems prostrate, spreading to form other colonies; simple, entire, linear basal leaves, five to ten in tufts; small white, sometimes tinted with pink, five-lobed tubular flowers borne singly on stalks (peduncles) shorter than leaves; fruit many-seeded round capsule.

Flowering period: June through September.

Habitat: Regularly flooded brackish and fresh tidal marshes and intertidal mud and sand flats.

Wetland indicator status: OBL.

Range: Newfoundland and Quebec south to North Carolina.

SEE ALSO Riverbank Quillwort (*Isoetes riparia*), Marsh Pennywort (*Hydrocotyle umbellata*), and Sweet Flag (*Acorus calamus*).

Yellow Flag

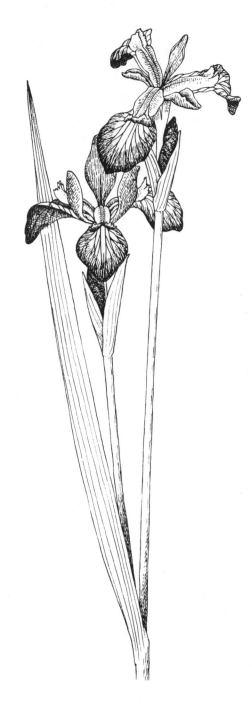

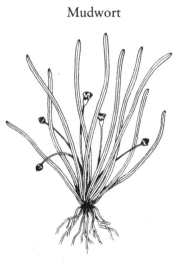

Mudwort

Northern Blue Flag

PRICKLY-STEMMED HERBS

Halberd-leaved Tearthumb

Polygonum arifolium L.
Buckwheat or Smartweed Family
Polygonaceae

Description: Reclining perennial herb, erect when young, up to 6 feet or longer; stem jointed, weak, several-angled, and prickly; simple, entire, hairy leaves (up to 8 inches long and to 5 inches wide), broadly arrowhead-shaped with triangular basal lobes, midrib prickly, alternately arranged; small pink flowers with four lobes in small, close clusters; lens-shaped dark brown to black nutlet (achene). (*Note:* Prickles on stem point downward.)

Flowering period: July through October.

Habitat: Tidal fresh marshes; nontidal marshes, wet meadows, and swamps.

Wetland indicator status: OBL.

Range: New Brunswick to Minnesota, south to Florida and Missouri.

Similar species: Arrow-leaved Tearthumb (*P. sagittatum*) has a four-angled stem, narrowly arrowhead-shaped leaves with rounded bases, and mostly pink but sometimes white and green flowers with five lobes; it is OBL.

Arrow-leaved Tearthumb

Polygonum sagittatum L.
Buckwheat or Smartweed Family
Polygonaceae

Description: Reclining perennial herb, erect when young, up to 6 feet or more long; stem jointed, weak, four-angled, and prickly; simple, entire leaves (⅕–4 inches long and about 1 inch wide), lance-shaped with arrowhead-shaped bases or narrowly arrowhead-shaped with somewhat rounded bases, midrib prickly, alternately arranged; small pink, sometimes white or green flowers with five petallike lobes on long stalked heads in leaf axils and terminally; three-angled dark brown or black nutlet (achene). (*Note:* Prickles on stem point downward.)

Flowering period: May through October.

Habitat: Tidal fresh marshes; nontidal marshes and wet meadows.

Wetland indicator status: OBL.

Range: Newfoundland and Quebec to Saskatchewan and Nebraska, south to Florida and Texas.

Similar species: Halberd-leaved Tearthumb (*P. arifolium*) is also prickly-stemmed, but has different leaves; see description and illustration.

SEE ALSO Dye Bedstraw (*Galium tinctorium*).

Halberd-leaved Tearthumb

Arrow-leaved Tearthumb

FLOWERING HERBS WITH SIMPLE, ENTIRE, ALTERNATE LEAVES

Marsh Spider Lily

Hymenocallis crassifolia Herbert

Amaryllis Family
Amaryllidaceae

Description: Medium-height, erect perennial herb, up to 2 feet tall, with conspicuous underground bulb; simple, entire, folded or creased, linear leaves (up to ½ inch wide) with tubular basal sheaths, alternately arranged; large showy, six-"petaled" white flowers (up to 4 inches long) with spreading funnel-shaped tubular center (corolla) borne singly or in twos (rarely in threes) in umbellike clusters; fruit capsules bearing one to three green, fleshy, roundish seeds.

Flowering period: Mid-May through June.

Habitat: Slightly brackish and tidal fresh marshes; nontidal marshes, muddy shores, and edges of forested wetlands.

Wetland indicator status: OBL.

Range: North Carolina south along the Coastal Plain to Florida, west to Alabama.

Similar species: Carolina Spider Lily (*H. caroliniana*, formerly *H. occidentalis*) occurs in similar habitats from Georgia south (rarely in North and South Carolina); it usually has three or more flowers (rarely two) per cluster; it is OBL. Southern Swamp Lily (*Crinum americanum*) has two or more (usually four) fragrant, six-petaled, white or pinkish flowers with a greenish, narrow tubular corolla per cluster; it occurs from Georgia to Florida and west to Texas; it is OBL.

Asiatic Dayflower

Murdannia keisak (Hassk.) Hand.-Mazz.
[*Aneilema keisak* Hassk.]

Spiderwort Family
Commelinaceae

Description: Low-growing, trailing annual herb, often forming dense mats; stems usually rooting at nodes; simple, entire, linear to narrowly lance-shaped leaves (up to 2¾ inches long and to ½ inch wide), somewhat clasping with somewhat hairy tubular sheaths, alternately arranged; pink to light purple, three-petaled flowers (about ½ inch long) with three projecting sepals, borne singly or in clusters (racemes) of two to four from leaf axils; two- or three-chambered, oval-shaped fruit capsules bearing several seeds.

Flowering period: September through October.

Habitat: Tidal fresh marshes; nontidal marshes, stream banks, shallow water, ditches, and edges of swamps.

Wetland indicator status: OBL.

Range: Maryland and southeastern Virginia to northern Florida; native of eastern Asia.

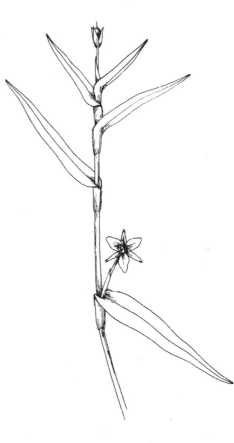

Asiatic Dayflower

Marsh Spider Lily

New York Aster

Aster novi-belgii L.

Composite or Aster Family
Compositae (Asteraceae)

Description: Medium-height to tall, erect perennial herb, up to 5 feet high; stems sometimes with lines of hairs from leaf bases, sometimes smooth, except under flower heads; simple, entire or weakly toothed, narrowly lance-shaped leaves (1½–6¾ inches long) slightly clasping stem, alternately arranged; violet or blue daisy-like flowers in heads (¾–1¼ inches wide) with twenty to fifty petallike rays (⅕–½ inch long) arranged in open or leafy inflorescences.

Flowering period: Late September into November.

Habitat: Slightly brackish and tidal fresh marshes, occasionally borders of salt marshes; nontidal marshes, shrub swamps, shores, pine barrens, and other moist areas.

Wetland indicator status: OBL.

Range: Newfoundland and Nova Scotia south to Georgia, apparently to Alabama, chiefly near the coast.

Similar species: Elliott's Aster (*A. elliottii*) occurs in brackish and tidal fresh marshes from southeastern Virginia to Florida (including the Panhandle); its primary leaves sheath the stem and its flowers are composed of purple-pink rays surrounding a yellow to red or purple central disk; it is OBL. Frost Aster (*A. pilosus*), a white-flowered aster, occurs in similar habitats from South Carolina north; it is FAC−.

Water Lotus

Nelumbo lutea (Willd.) Pers.

Water Lily Family
Nymphaeaceae

Description: Aquatic to emergent perennial herb, up to 3½ feet tall; rhizome spongy; simple, entire, rounded leaves (up to 2 feet wide) on stalks (up to 3½ feet long) attached to center of underside, alternately arranged; large showy, many-petaled yellowish flowers (up to 10 inches wide) borne singly on long stalks; many acorn-like nutlets (about ½ inch wide) borne in dark brown, cuplike fruit capsule (up to 4 inches wide).

Flowering period: June through September.

Habitat: Slightly brackish and fresh tidal marshes and waters; shallows of ponds and streams.

Wetland indicator status: OBL.

Range: Southern Ontario and Massachusetts to Florida, west to eastern Texas and Minnesota.

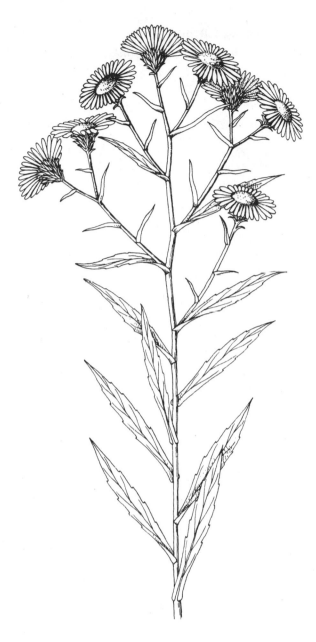

Water Lotus flower

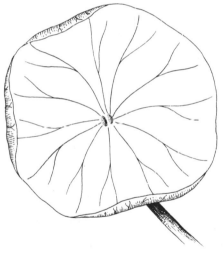

New York Aster

Water Lotus leaf

Bushy Seedbox

Ludwigia alternifolia L.

Evening Primrose Family
Onagraceae

Description: Medium-height, erect perennial herb, 1½–3½ feet tall; simple, entire, sessile leaves (2–4 inches long), lance-shaped and pointed, wedge-shaped or tapered bases, alternately arranged; small four-petaled yellow flowers (to 1 inch wide) borne singly in leaf axils; fruit capsule square at top, rounded at base, with a terminal pore.

Flowering period: May through October.

Habitat: Tidal fresh marshes; nontidal marshes, wet meadows, swamps, and wet soils.

Wetland indicator status: OBL.

Range: Massachusetts and southern Ontario to Iowa and Kansas, south to Florida and Texas.

Similar species: Other *Ludwigia* with four stamens found in coastal wetlands have blunt or rounded leaf bases: Seaside Seedbox (*L. maritima*) has stalked flowers and fruit capsules; it is FACW. Winged Seedbox (*L. alata*) has sessile flowers lacking petals, and capsules with wings; it is OBL. Other erect coastal marsh *Ludwigia* have yellow flowers with five to seven petals and eight or more stamens: River Seedbox (*L. leptocarpa*) has short-stalked (¾ inch or less), five- to seven-petaled flowers and elongate, many-ribbed, long-hairy fruit capsules; it is OBL. Uruguay Seedbox (*L. uruguayensis*) has somewhat longer flower stalks (½–2 inches long) and often forms extensive mats; it is OBL. Primrose-willow (*L. peruviana*) may occur in tidal fresh marshes in central and southern Florida; it grows to about 10 feet tall from woody lower stems; it has long leaves (3½–6 inches long), and large four- or five-petaled yellow flowers (up to 2½ inches wide); it is OBL.

Spring Ladies'-tresses

Spiranthes vernalis Engelm. & Gray

Orchid Family
Orchidaceae

Description: Medium-height, erect perennial herb, up to 3½ feet tall; roots fleshy and long; simple, entire, linear, grasslike, mostly basal leaves (up to 12 inches long and about ½ inch wide) often ridged or somewhat rounded in cross-section, sheathing at base, smaller alternately arranged leaves sometimes present on flowering stalk; many fragrant, yellowish, greenish, or whitish, two-lipped, somewhat fleshy flowers (about ½ inch long) borne on upper part of separate flowering stem (scape) forming a dense, one-sided, sometimes twisted spike (up to 10 inches long) covered with brownish hairs.

Flowering period: May into October.

Habitat: Tidal fresh marshes; nontidal marshes, wet meadows, bogs, dry or moist open fields, and swales.

Wetland indicator status: FACW−.

Range: Southern New England and New York to eastern Kansas, south to Florida and eastern Texas; also in Central America.

Similar species: Marsh Ladies'-tresses (*S. odorata*) occurs in brackish and tidal fresh marshes from Maryland south; its flowers face two to four directions and bloom from September to December; it is OBL.

Bushy Seedbox flower

Bushy Seedbox

River Seedbox
flower and fruit

Spring Ladies'-tresses

Pinkweed
or **Pennsylvania Smartweed**

Polygonum pensylvanicum L.

Buckwheat or Smartweed Family
Polygonaceae

Description: Medium-height to tall annual herb, up to 6½ feet high; jointed stems, upper stem covered with glandular hairs; simple, entire, lance-shaped leaves (up to 10 inches long) usually hairy below, stalked, alternately arranged, leaf sheaths (ocrea) without hairy fringe; many small five-lobed, pink or purplish flowers (⅛ inch long) borne in dense clusters on spikelike terminal inflorescences; lens-shaped nutlets.

Flowering period: May through October.

Habitat: Tidal fresh marshes; nontidal marshes, wet meadows, damp fields, cropland, and gardens.

Wetland indicator status: FACW.

Range: Nova Scotia and Quebec to Minnesota and South Dakota, south to Florida and Texas.

Similar species: Mild Water Pepper (*P. hydropiperoides*) may also have pink or purplish flowers, but they are arranged in loose spikes; its leaf sheaths are fringed with hairs; it is OBL. Nodding Smartweed (*P. lapathifolium*) has its flowers borne in dense nodding spikes; it is FACW. See also Water Smartweed (*P. amphibium*).

Dotted Smartweed

Polygonum punctatum Elliott

Buckwheat or Smartweed Family
Polygonaceae

Description: Medium-height, erect annual herb, up to 3½ feet tall; stems jointed and sheathed above each joint; simple, entire, smooth leaves (up to 8 inches long and usually less than ½ inch wide); lance-shaped, tapering at both ends, alternately arranged, leaf sheaths (ocrea) smooth or few-haired; numerous small green or greenish white flowers arranged in loose, erect spikes, calyx dotted with glands; lens-shaped or three-sided shiny nutlets.

Flowering period: June to October.

Habitat: Tidal fresh marshes (regularly and irregularly flooded zones), occasionally slightly brackish marshes; nontidal marshes, wet soils, open swamps, and shallow waters.

Wetland indicator status: FACW+.

Range: Quebec to British Columbia, south to Florida and California.

Similar species: Common Smartweed (*P. hydropiper*) has greenish or red-tipped flowers in loose spikes curving at tips, calyx dotted with glands, and dull lens-shaped or three-sided nutlet; it is OBL. Mild Water Pepper (*P. hydropiperoides*) has mostly pink or purplish flowers, sometimes white or green, in erect, loose cylinder-shaped spikes, and its fruit has a mild peppery taste; also, its calyx is not dotted with glands; it is OBL. Swamp Smartweed (*P. setaceum*) has mostly white or greenish flowers and closely resembles *P. hydropiperoides*, but has wider leaves (more than ½ inch wide) and leaf sheaths covered with long spreading hairs (as opposed to short appressed hairs); it is FACW. Pinkweed (*P. pensylvanicum*) has pinkish or purplish flowers in dense erect spikes; it is FACW.

Pinkweed

Dotted Smartweed
flower

Dotted Smartweed

Virginia Knotweed or Jumpseed

Polygonum virginianum L.
[*Tovara virginiana* (L) Raf.]

Buckwheat or Smartweed Family
Polygonaceae

Description: Medium-height perennial herb, up to 4 feet tall; jointed stems; simple, entire, fine-pointed, egg-shaped leaves (up to 6 inches long) on short stalks, alternately arranged, leaf sheaths (ocrea) fringed with hairy bristles; leafless terminal inflorescence (up to 20 inches long) bearing numerous widely spaced clusters of one to three very small greenish white (sometimes pinkish) flowers.

Flowering period: July into November.

Habitat: Tidal swamps; nontidal temporarily flooded forested wetlands, and moist woods and thickets.

Wetland indicator status: FAC.

Range: New Hampshire to Minnesota and Nebraska, south to Florida and Texas.

Swamp Dock

Rumex verticillatus L.

Buckwheat or Smartweed Family
Polygonaceae

Description: Medium-height, erect annual herb, up to 3½ feet tall; stems jointed and grooved; simple, entire, pale green, narrowly lance-shaped flat leaves tapering at base to petiole, alternately arranged; numerous small green flowers, often tinged with red, borne singly on long drooping stalks (pedicels, ⅖–⅗ inch long) arranged in whorls along inflorescence (1–1½ inches long); three-winged fruit.

Flowering period: April to September.

Habitat: Tidal fresh marshes; nontidal marshes and swamps, and edges of streams.

Wetland indicator status: FACW+.

Range: Quebec and Ontario to Wisconsin and Kansas, south to Florida and Texas.

Similar species: Curly Dock (*R. crispus*), an introduction from Eurasia, has dark green leaves with curled margins; it is FAC. Bitter Dock (*R. obtusifolius*) also has dark green leaves with wavy margins, but its leaf veins are red; it is FACW−.

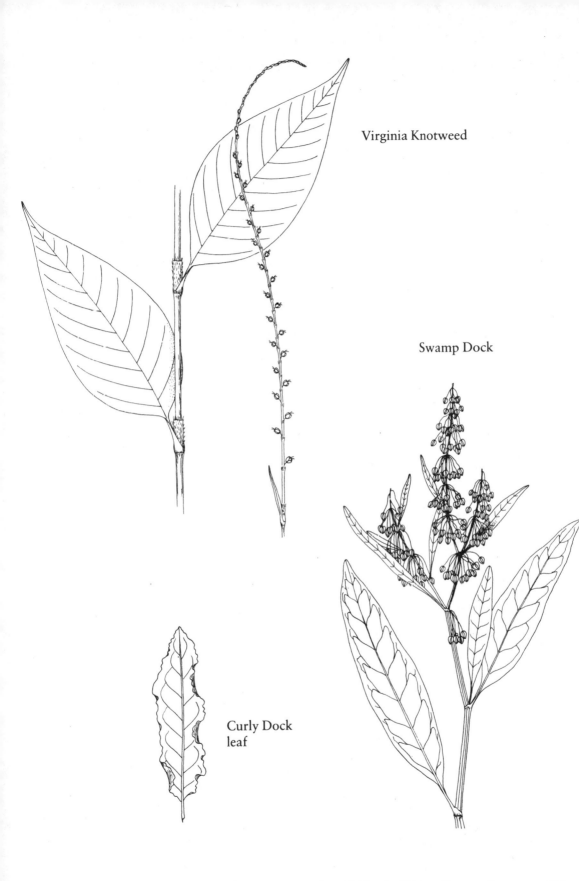

Virginia Knotweed

Swamp Dock

Curly Dock
leaf

Water Pimpernel

Samolus parviflorus Raf.
[*Samolus floribundus* H.B.K.]

Primrose Family
Primulaceae

Description: Low to medium-height, erect perennial herb, 4–24 inches tall; stems branched from upper half and also from base; simple, entire, spoon-shaped or somewhat oval leaves (mostly 1–2 inches long, sometimes to 5 inches), both basal and alternately arranged; small white, five-lobed, bell-shaped flowers on slender spreading stalks (pedicels, up to ⅕ inch long) with small bract near middle of stalk, borne in terminal inflorescences (racemes, 1¼–6 inches long); round fruit capsule bearing many seeds.

Flowering period: March to September.

Habitat: Brackish and tidal fresh marshes (regularly and irregularly flooded zones) and shallow tidal waters; inland sandy and muddy stream banks, lake shores, and ditches.

Wetland indicator status: OBL.

Range: New Brunswick to southern Michigan, Missouri, and Kansas, south to Florida and Texas; also in western states.

Similar species: Coast Water Pimpernel (*S. ebracteatus*) has pinkish flowers, and its flower stalks lack bracts; it occurs in salt and brackish marshes, edges of mangrove swamps, dunes, and wet sands along the Gulf coast; it is OBL.

Lizard's Tail

Saururus cernuus L.

Lizard's Tail Family
Saururaceae

Description: Medium-height to tall, erect perennial herb, up to 4 feet high; stems jointed and slightly branching; simple, entire, somewhat heart-shaped, broad leaves (2¼–6 inches long and to 3 inches wide) tapering to a point distally, borne on long petioles and sheathing stem at base, alternately arranged; numerous small white, fragrant flowers borne on one or two slender terminal spikes (up to 8 inches long), nodding at tip before all flowers mature; somewhat rounded, wrinkled fruit capsule.

Flowering period: May through July.

Habitat: Tidal fresh marshes and swamps (regularly and irregularly flooded zones); nontidal swamps, marshes, and shallow waters.

Wetland indicator status: OBL.

Range: Southern New England, southern Quebec, and Minnesota south to Florida and Texas.

SEE ALSO Southern Water Plantain (*Alisma subcordatum*).

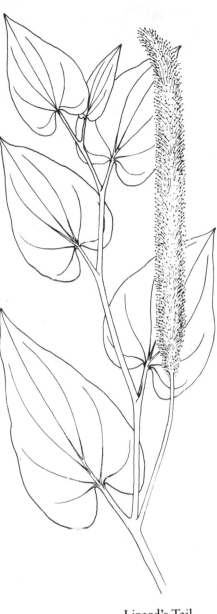

Water Pimpernel

Lizard's Tail

FLOWERING HERBS WITH SIMPLE, ENTIRE, OPPOSITE LEAVES

Swamp Milkweed

Asclepias incarnata L.

Milkweed Family
Asclepiadaceae

Description: Medium-height to tall, erect perennial herb, up to 6 feet high; stems round and smooth or hairy, with milky sap; simple, entire, lance-shaped leaves, rounded or tapering at base, smooth or hairy on both sides, oppositely arranged; numerous pink to purplish red regular flowers composed of five erect hoods (⅓ inch long) with somewhat longer horns and five downward-pointing lobes, arranged in several umbels; fruit pod tapered at both ends and standing erect from branches.

Flowering period: June into September.

Habitat: Tidal fresh marshes and edges of brackish marshes; nontidal shrub swamps, forested wetlands, marshes, shores, and ditches.

Wetland indicator status: OBL.

Range: Nova Scotia to Manitoba to Utah, south to Florida, Louisiana, and New Mexico.

Similar species: Red Milkweed (*A. lanceolata*) occurs in brackish and fresh marshes from southern New Jersey south; its leaves are linear or narrowly lance-shaped, and its flowers are orange or red; it is OBL.

Dwarf St. John's-wort

Hypericum mutilum L.

St. John's-wort Family
Guttiferae (*Hypericaceae*)

Description: Low to medium-height, erect perennial herb, up to 28 inches tall; simple, entire, oval to somewhat egg-shaped or lance-shaped leaves (up to 2 inches long and to ¾ inch wide) with pointed tips, clasping bases, dark spots on lower or both leaf surfaces, stalkless, lower stem leaves much smaller than upper leaves, oppositely arranged; many small five-petaled yellow flowers (less than ⅕ inch wide) borne in open, branched inflorescence; elongate fruit capsule (less than ⅕ inch wide) bearing minute seeds.

Flowering period: June through October.

Habitat: Tidal fresh marshes; nontidal marshes, wet meadows, bogs, forested wetlands, stream banks, river bars, swales, wet or moist open soil, and ditches.

Wetland indicator status: FACW.

Range: Newfoundland to Manitoba, south to southern Florida and eastern Texas.

Similar species: Sandweed St. John's-wort (*H. fasciculatum*) is a low shrub (up to 7 feet tall) with very narrow linear leaves; it is FACW+. Marsh St. John's-wort (*Triadenum virginicum*) has reddish, purplish, or pinkish flowers; it is OBL.

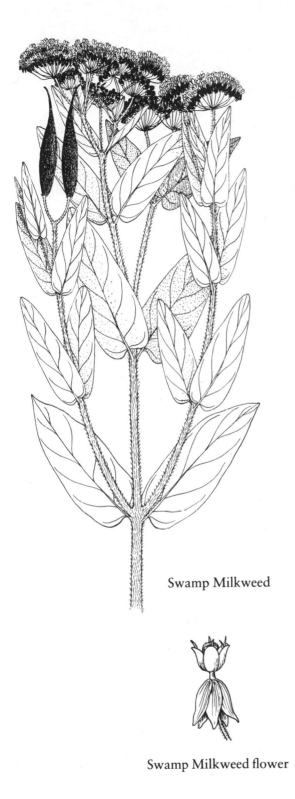

Swamp Milkweed

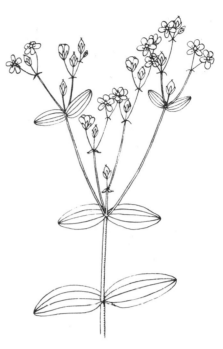

Dwarf St. John's-wort

Dwarf St. John's-wort habit

Swamp Milkweed flower

Marsh St. John's-wort

Triadenum virginicum (L.) Raf.
[*Hypericum virginicum* L.]

St. John's-wort Family
Guttiferae (Hypericaceae)

Description: Medium-height, erect, perennial herb, 1–2½ feet tall; stems unbranched, commonly reddish; simple, entire, oblong to egg-shaped leaves (1⅕–2⅖ inches long), usually with round tips and somewhat heart-shaped bases, sessile, marked with translucent dots (glands) beneath, oppositely arranged; numerous five-petaled reddish, pinkish, or purplish flowers (up to ⅘ inch in diameter) borne on terminal and axillary inflorescences; fruit tapered capsule.

Flowering period: July through September.

Habitat: Tidal fresh marshes (regularly and irregularly flooded zone), occasionally slightly brackish marshes; nontidal marshes, bogs, swamps, and wet shores.

Wetland indicator status: OBL.

Range: Nova Scotia to Florida and Mississippi; inland, New York to southern Ontario and Illinois.

Similar species: Dwarf St. John's-wort (*H. mutilum*) also occurs in tidal fresh marshes; its flowers are yellow and much smaller (less than ⅕ inch wide); it is FACW.

Pink Ammania

Ammania latifolia L.
[*Ammania teres* Raf.]

Loosestrife Family
Lythraceae

Description: Low to medium-height, erect annual herb, up to 3 feet tall; somewhat fleshy stems round to somewhat triangular; simple, entire, linear to somewhat lance-shaped or spoon-shaped leaves (up to 2½ inches long) stalkless or short-stalked, upper leaves clasping stem, lower leaves tapered at base, oppositely arranged; small purplish, pinkish, or whitish four-"petaled" tubular flowers (less than ⅓ inch long) borne in leaf axils on short-stalked clusters (cymes) of two to ten flowers; purplish or reddish brown, round fruit capsules (about ⅕ inch wide) bearing shiny yellow seeds.

Flowering period: July through September.

Habitat: Tidal fresh and brackish marshes; nontidal marshes, ditches, and open edges of forested wetlands.

Wetland indicator status: OBL.

Range: New Jersey south along the Coastal Plain to Florida, west to eastern Texas; also in tropical America.

Similar species: Purple Ammania (*A. coccinea*) has larger leaves (to 6½ inches long) that clasp or nearly clasp the stem; it is FACW+. Toothcup (*Rotala ramosior*) lacks clasping leaves; it usually has only one flower per leaf axil, and its yellow fruit capsule has minute lines representing internal partitions; it is OBL. Buttonweed (*Diodia virginiana*) has usually one showy white hairy flower (about ¼ inch wide) in the leaf axils and often reddish-tinged stems; it is FACW.

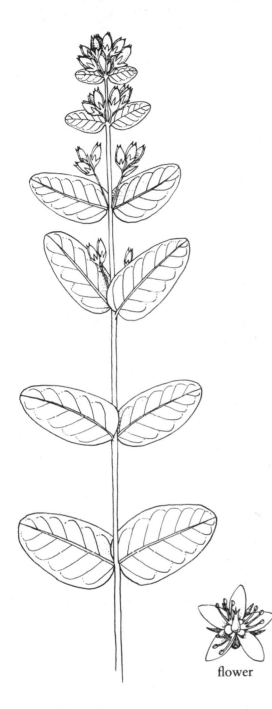

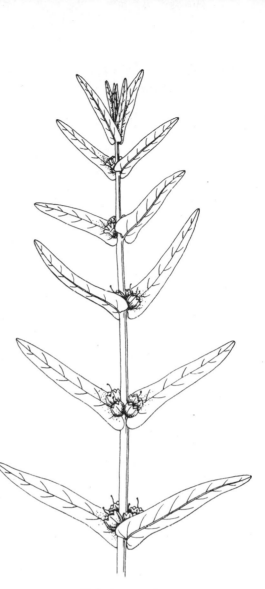

Pink Ammania

flower

Marsh St. John's-wort

Water or Marsh Purslane

Ludwigia palustris (L.) Elliott

Evening Primrose Family
Onagraceae

Description: Creeping, mat-forming or floating perennial herb; stems smooth, soft, weak, and commonly reddish; simple, entire leaves (up to 1¼ inches long), oppositely arranged; very small greenish flowers (lacking petals or with four reddish petals when growing out of water) borne singly in leaf axils; four-sided fruit capsules (less than ⅕ inch wide).

Flowering period: May into October.

Habitat: Tidal fresh marshes and waters; shallow water, nontidal marshes, muddy shores, and wet soil.

Wetland indicator status: OBL.

Range: Nova Scotia to Oregon, south to Florida and Mexico.

Similar species: Two related mat-forming or floating species also occur in similar habitats: Floating Seedbox (*L. peploides*) and Uruguay Seedbox (*L. uruguayensis*). Both have five- or six-petaled yellow flowers. The former usually has smooth, erect flowering stems, whereas the latter has floating or creeping flowering stems, often covered by long, soft hairs. Both are OBL.

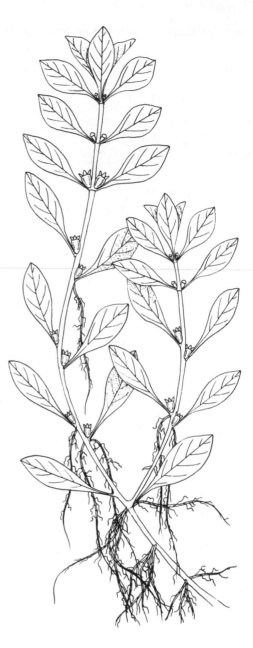

Water Purslane

SEE ALSO Alligatorweed (*Alternanthera philoxeroides*), Water-willow or Swamp Loose-strife (*Decodon verticillatus*), False Pimpernel (*Lindernia dubia*), described under Overlooked Hedge Hyssop (*Gratiola neglecta*), and Purple Gerardia (*Agalinis purpurea*) described under Seaside Gerardia (*A. maritima*).

FLOWERING HERBS WITH SIMPLE, ENTIRE, WHORLED LEAVES

Water-willow
or Swamp Loosestrife

Decodon verticillatus (L.) Elliott

Loosestrife Family
Lythraceae

Description: Medium-height, erect herb with somewhat woody base and distinctly arching branches up to 9 feet long and often rooting at tips; stems four- to six-angled; simple, entire, narrowly lance-shaped leaves (2–6 inches long), fine-pointed, short-petioled, oppositely arranged or in whorls of threes or fours; numerous five-petaled, pink bell-shaped flowers on short stalks borne in clusters in axils of upper leaves; round-ish fruit capsule (about ⅕ inch wide) with three to five cells.

Flowering period: July into October.

Habitat: Regularly flooded tidal fresh marshes; borders of rivers, ponds, and lakes, nontidal marshes, shrub swamps, and forested wetlands.

Wetland indicator status: OBL.

Range: Central Maine and southern New Hampshire to southern Ontario and Illinois, south to Florida, Louisiana, and Texas.

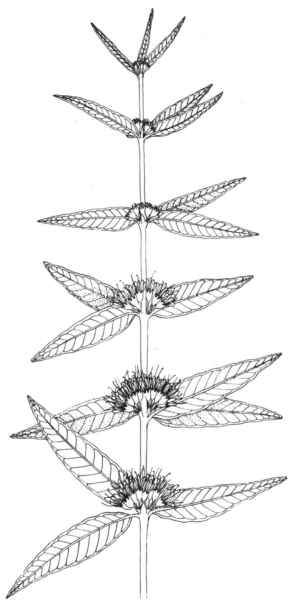

flower

Water-willow

Dye Bedstraw
or Clayton's Bedstraw
Galium tinctorium L.

Madder Family
Rubiaceae

Description: Low to medium-height, weakly erect or matted herb, up to 2 feet tall or long; stems four-angled, with sharp recurved teeth or prickles on angles, and much branched; simple, entire, oblong to lance-shaped leaves (⅕–⅖ inch long) with rough margins and midveins, arranged in whorls of usually five or six on main stem and of two to four on branches, leaves of single whorl often differing in size; small white three-lobed flowers with smooth, short, and straight stalks (pedicels) borne in clusters of threes; small black, roundish, dry fruits.

Flowering period: April through June.

Habitat: Irregularly flooded tidal fresh marshes; nontidal marshes, bogs, and swamps.

Wetland indicator status: FACW.

Range: Quebec and Newfoundland to Washington, south to North Carolina, Texas, and Arizona.

Dye Bedstraw

FLOWERING HERBS WITH SIMPLE, TOOTHED, ALTERNATE LEAVES

Jewelweed
or **Spotted Touch-me-not**

Impatiens capensis Meerb.

Touch-me-not Family
Balsaminaceae

Description: Medium-height to tall, erect annual herb, 2–5 feet high, rarely 6 feet; stems smooth and somewhat succulent; simple, coarsely toothed, soft (almost fleshy), egg-shaped leaves (1⅕–4 inches long) on petioles, alternately arranged; few to several orange or orange-yellow, seemingly three-petaled tubular flowers (⅘–1⅕ inches long) with reddish brown spots and curved spur at end, borne on long, drooping axillary stalks (pedicels); fruit capsulelike (about ⅘ inch long). (*Note:* Plants vary in height, leaf size, and color of foliage with differences in exposure and available moisture.)

Flowering period: May into November.

Habitat: Tidal fresh marshes, occasionally slightly brackish marshes; nontidal marshes and swamps, moist woods, stream banks, and springs.

Wetland indicator status: FACW.

Range: Newfoundland and Quebec west to Saskatchewan, south to Florida, Alabama, Arkansas, and Oklahoma.

Jewelweed

Cardinal Flower

Lobelia cardinalis L.

Bluebell Family
Campanulaceae

Description: Medium-height to tall, erect perennial herb, 1½–5 feet high; stem typically unbranched, smooth or slightly hairy; simple, fine- or round-toothed, lance-shaped to oblong leaves (2–6 inches long and up to 2 inches wide) tapering at both ends, smooth or fine-hairy, lower leaves short-stalked, upper leaves sessile, alternately arranged; numerous bright red (sometimes white) two-lipped tubular flowers (up to 1⅗ inches long) borne on terminal spikelike inflorescences (racemes, up to 20 inches long), sometimes in leaf axils, upper lip of flower two-lobed and erect, lower lip three-lobed and spreading downward; fruit two-celled capsule.

Flowering period: May into December.

Habitat: Irregularly flooded tidal fresh marshes and swamps; nontidal marshes, wet meadows, swamps, springs, and riverbanks.

Wetland indicator status: FACW+.

Range: New Brunswick to Michigan and Minnesota, south into Florida and Texas, west to Nevada and California; also in Mexico.

Similar species: Elongated Lobelia (*L. elongata*) has bluish or purplish flowers and thick, sharp-toothed, almost spiny leaves; it is OBL.

Elongated Lobelia

Lobelia elongata Small

Bluebell Family
Campanulaceae

Description: Medium-height to tall, erect perennial herb, up to 6 feet high; stem smooth; simple, sharp-toothed (sometimes round-toothed), thick or somewhat fleshy, linear to lance-shaped leaves (up to 6 inches long and to 1¾ inches wide), stalked (often by "wings" of leaf blades), alternately arranged; many showy, bluish or purplish two-lipped tubular flowers (about 1 inch long) borne on one side of terminal inflorescence (raceme, to 20 inches long), subtended by leaflike bracts, flower upper lip two-lobed and lower lip three-lobed; fruit capsule (about ½ inch wide).

Flowering period: August through October.

Habitat: Brackish and tidal fresh marshes and swamps; nontidal marshes, bogs, swamps, and savannahs.

Wetland indicator status: OBL.

Range: Delaware and Maryland south to Georgia.

Elongated Lobelia

Elongated Lobelia flower

Cardinal Flower

Small White Aster

Aster vimineus Lam.

Composite or Aster Family
Compositae (*Asteraceae*)

Description: Medium-height to tall perennial herb, up to 5 feet high; smooth, usually purplish, very leafy stems, sometimes with hairs arranged in lines; simple, toothed or entire, narrowly lance-shaped leaves (up to 4½ inches long along stem and much reduced on upper branches), smooth below, alternately arranged; many small, white (rarely purplish), daisylike flowers (up to ½ inch wide) with fifteen to thirty petallike rays surrounding a yellow, red, or purplish central disk, borne on leafy branches from upper leaf axils.

Flowering period: September into October.

Habitat: Fresh tidal marshes and swamps; nontidal marshes, wet meadows, floodplains, and other moist areas.

Wetland indicator status: FAC.

Range: Maine south along the Coastal Plain to northern Florida, west to Louisiana and north in Mississippi River valley to southern Ohio and western Missouri.

Similar species: Calico Aster (*A. lateriflorus*) has leaves that are hairy below, at least on the veins, and its flowers have nine to twenty white (rarely light purplish) petallike rays; it is FAC. Panicled or White Lowland Aster (*A. simplex*) has larger flowers (¾–1 inch wide) and is not as leafy; it is FACW.

White Boltonia

Boltonia asteroides (L.) L'Her.

Composite or Aster Family
Compositae (*Asteraceae*)

Description: Tall, erect perennial herb, up to 6½ feet high; rhizomes long and slender, simple, coarse-toothed, somewhat leathery, egg-shaped to linear leaves (to 10 inches long and to 1⅓ inches wide) with pointed tips, stalkless or nearly so, alternately arranged; few to many daisylike flowers composed of twenty-five to thirty-five white, pink, or lavender rays (to ⅘ inch long) surrounding a yellow central disk; flattened two-awned nutlets.

Flowering period: August through October.

Habitat: Slightly brackish and tidal fresh marshes; nontidal marshes, savannahs, muddy shores, and ditches.

Wetland indicator status: FACW.

Range: Southern New Jersey to Florida, west to Texas.

Similar species: Carolina Boltonia (*B. caroliniana*) lacks rhizomes and grows to more than 9 feet tall; it is FACW.

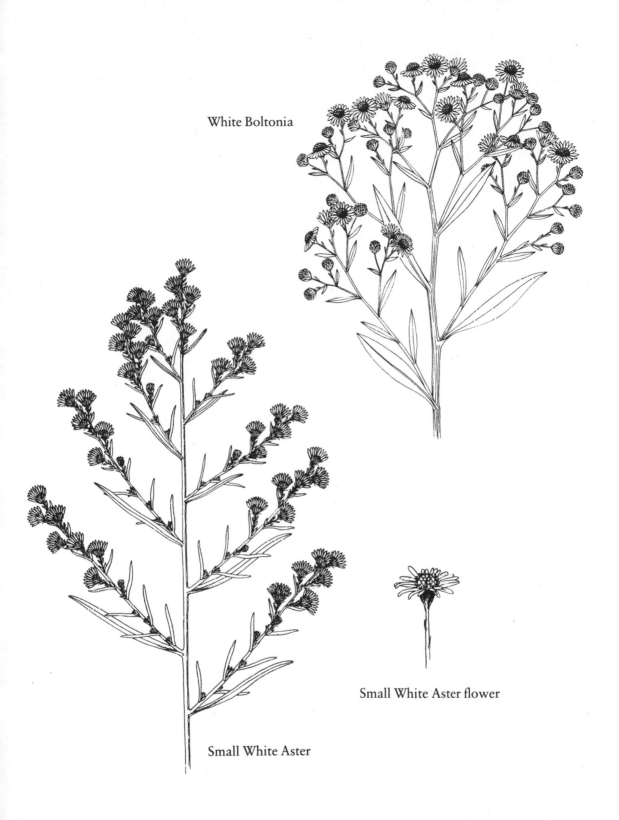

White Boltonia

Small White Aster flower

Small White Aster

New York Ironweed

Vernonia noveboracensis (L.) Michx.
Composite or Aster Family
Compositae (Asteraceae)

Description: Medium-height to tall, erect perennial herb, 3–7 feet high; stems smooth or thinly hairy; simple, fine-toothed or nearly entire, lance-shaped or narrowly lance-shaped leaves (4–8 inches long), rough-hairy above and thin-hairy beneath, alternately arranged; twenty-nine to forty-seven purple flowers in heads (½–¾ inch wide) arranged in loose, open, flattened, or round-topped inflorescence.

Flowering period: July into October.

Habitat: Tidal fresh marshes; nontidal swamps, marshes, wet meadows, and stream banks, mostly near the coast.

Wetland indicator status: FAC+.

Range: Massachusetts to Georgia and Mississippi, occasionally inland to Ohio and West Virginia.

Marsh Eryngo

Eryngium aquaticum L.

Parsley Family
Umbelliferae (Apiaceae)

Description: Tall perennial or biennial herb, up to 6½ feet high; stems parallel-ribbed; simple, toothed or wavy-margined, linear to somewhat lance-shaped basal leaves (6–20 inches long and to 3⅕ inches wide), broadest near tip, upper stem leaves weakly toothed, alternately arranged; somewhat bluish flowers in many roundish heads (about ½ inch wide) subtended by bluish, spiny, leaflike bracts (much longer than flowering heads) forming open axillary and terminal inflorescences.

Flowering period: July through September.

Habitat: Brackish and tidal fresh marshes; nontidal swamps, wet pinelands, bogs, ponds, streams, and ditches.

Wetland indicator status: OBL.

Range: New Jersey south along the Coastal Plain to central Florida, west to Texas.

Similar species: Rattlesnake-master (*E. yuccifolium*) has whitish or greenish flowers lacking spiny leaflike bracts below; it is FAC.

SEE ALSO Rose Mallow (*Hibiscus moscheutos*) and New York Aster (*Aster novi-belgii*).

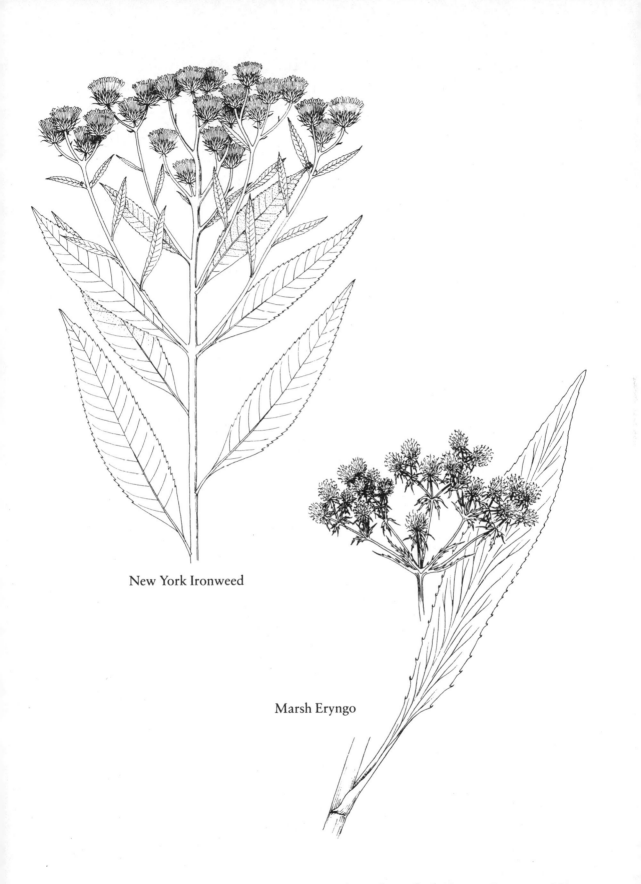

New York Ironweed

Marsh Eryngo

FLOWERING HERBS WITH SIMPLE, TOOTHED, OPPOSITE LEAVES

Giant Ragweed

Ambrosia trifida L.

Composite or Aster Family
Compositae (*Asteraceae*)

Description: Medium-height to tall, erect annual herb, up to 17 feet high (occasionally); stems hairy above and smooth below; simple, sharply three- to five-lobed, toothed leaves (up to 8 inches long), rough on both sides, oppositely arranged; small green flowers in heads on dense spikes.

Flowering period: Late June into October.

Habitat: Tidal fresh marshes; riverbanks, moist soils, and waste places.

Wetland indicator status: FAC.

Range: Quebec to British Columbia, south to Florida and northern Mexico.

Bur Marigold

Bidens laevis (L.) B.S.P.

Composite or Aster Family
Compositae (*Asteraceae*)

Description: Medium-height, erect annual or perennial herb, 1–3½ feet tall; stems smooth; simple, toothed, sessile, lance-shaped leaves (2–6 inches long), midrib prominent, oppositely arranged; yellow daisylike flowers in small heads (1½–2½ inches wide) with seven to eight petal-like yellow rays (⅘–1⅓ inches long); barbed nutlet (achene).

Flowering period: August to November.

Habitat: Tidal fresh marshes; nontidal marshes and borders of ponds and streams.

Wetland indicator status: OBL.

Range: New Hampshire and Massachusetts to Florida and California, chiefly coastal but also inland in the North to Indiana and West Virginia.

Similar species: Large-fruit Beggar-ticks (*B. coronata*), Small-fruit Beggar-ticks (*B. mitis*), and Devil's Beggar-ticks (*B. frondosa*) occur in similar habitats, with the first two also found in brackish marshes. They all have compound leaves, but the first two have leaflets that are also deeply lobed or divided. *B. coronata* and *B. mitis* are similar OBL species, having flowers with eight yellow petals (rays) or petalless, whereas *B. frondosa* is petalless and FACW. *B. coronata* has leaflike bracts (below flowers) with a brown to purple stripe in center and yellow margins and nutlets longer than ⅓ inch. *B. mitis* has bracts with a brown stripe and several brown lines in center and yellow margins and nutlets ⅓ inch long or less.

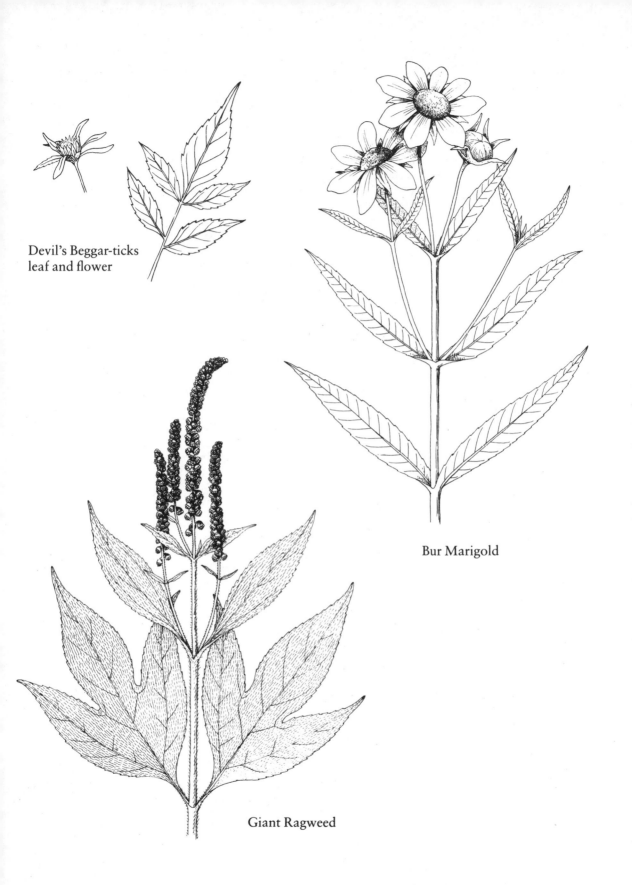

Devil's Beggar-ticks
leaf and flower

Bur Marigold

Giant Ragweed

Boneset

Eupatorium perfoliatum L.

Composite or Aster Family
Compositae (Asteraceae)

Description: Medium-height to tall, erect perennial herb, 1½–5 feet high; stems with long spreading hairs, occasionally densely hairy or coarsely hairy; simple, coarsely toothed, triangle-shaped leaves (2¾–8 inches long) joined at bases to form a single leaf, oppositely arranged; nine to twenty-three white flowers in heads borne on a flat-topped inflorescence (corymb, to 16 inches wide).

Flowering period: Late July through October.

Habitat: Tidal fresh marshes; nontidal marshes, wet meadows, shrub swamps, low woods, shores, and other moist areas.

Wetland indicator status: FACW+.

Range: Nova Scotia and Quebec to Minnesota and Nebraska, south to Florida, Louisiana, and Texas.

Similar species: Late-flowering Thoroughwort or Boneset (*E. serotinum*) has white flowers and petioled leaves; it occurs in brackish and tidal fresh marshes; it is FAC. Mistflower (*Conoclinium coelestinum*, formerly *E. coelestinum*) has violet or light purple flowers and petioled leaves somewhat triangle-shaped and toothed; it is FAC. Dog-fennel (*E. capillifolium*), an aggressive weedy relative of *E. perfoliatum*, occurs on dikes surrounding impounded freshwater marshes along tidal rivers; it has compound leaves divided into many linear to threadlike leaflets (lower leaves oppositely arranged, upper leaves alternately arranged) and it bears many small creamy white flowers in heads borne on a much-branched, somewhat narrowly pyramid-shaped inflorescence (panicle); it is FACU.

Water Horehound or Bugleweed

Lycopus virginicus L.

Mint Family
Labiatae (Lamiaceae)

Description: Medium-height, erect perennial herb, up to 3 feet tall, usually from tuberous root; stems four-angled with rounded edges, usually fine-hairy, often producing long runners from base; simple, coarse-toothed, lance-shaped leaves (2–5 inches long) tapered at both ends, with coarse marginal teeth beginning just below middle of leaf, leaves generally dark green, sometimes purple-tinged, oppositely arranged; numerous small four-petaled white tubular flowers borne in dense ball-like clusters at leaf bases; nutlet longer than persistent sepals.

Flowering period: June to October.

Habitat: Irregularly flooded tidal fresh marshes; nontidal marshes, wet meadows, and forested wetlands.

Wetland indicator status: OBL.

Range: Quebec and Nova Scotia to Minnesota, south to northern Florida and eastern Texas.

Similar species: Tapertip Bugleweed (*L. rubellus*) is quite similar, but its mature nutlets are shorter than the persistent sepals; it is OBL. American Water Horehound (*L. americanus*) also occurs in tidal fresh marshes; its lower leaves are deeply lobed, and it lacks a tuberous root; it is OBL.

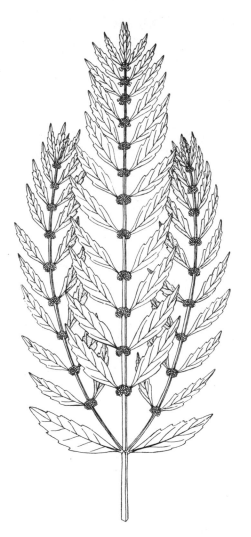

Water Horehound

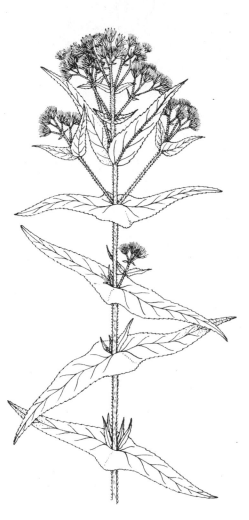

Boneset

American Water Horehound lower leaves

Mad-dog Skullcap

Scutellaria lateriflora L.

Mint Family
Labiatae (*Lamiaceae*)

Description: Medium-height, erect perennial herb, from 1–2½ feet tall; rhizomes slender; stems four-angled, slender, usually branched, smooth or fine-hairy on angles; simple, coarse-toothed, broadly lance-shaped leaves (up to 3 inches long) tapering to a point distally, rounded at base, petioled, oppositely arranged; numerous small blue, sometimes pink or white, two-lipped tubular flowers borne on one side of inflorescences (racemes) in leaf axils and usually one terminal, flowers usually subtended by small lance-shaped leaves.

Flowering period: July into October.

Habitat: Irregularly flooded tidal fresh marshes; nontidal marshes, wet meadows, and swamps.

Wetland indicator status: FACW+.

Range: Quebec to British Columbia, south to Florida, Louisiana, and Arizona.

Similar species: Common Skullcap (*S. galericulata*, formerly *S. epilobiifolia*) has blue flowers borne singly from leaf axils and short-petioled or stalkless leaves; it is OBL.

Overlooked Hedge Hyssop or Clammy Hedge Hyssop

Gratiola neglecta Torr.

Figwort Family
Scrophulariaceae

Description: Low-growing, erect annual herb, 4–12 inches tall; stem unbranched or widely branched, upper part hairy and sometimes sticky; simple, coarse, shallow-toothed, lance-shaped to oblong lance-shaped leaves (⅘–2½ inches long) tapering at both ends, most leaves sessile but young ones stalked, oppositely arranged; small yellow or cream-colored tubular flowers (to about ½ inch long) with five white lobes borne on somewhat drooping slender stalks (up to 1 inch long); roundish fruit capsule.

Flowering period: May to October.

Habitat: Regularly flooded mud flats along edges of tidal fresh marshes; inland shores and shallow waters.

Wetland indicator status: OBL.

Range: Quebec to southern British Columbia, south to Georgia, Texas, and Arizona.

Similar species: Virginia Hedge Hyssop (*G. virginiana*) has white tubular flowers internally marked with purple lines borne on short erect stalks; it is OBL. Golden-pert (*G. aurea*) has bright yellow tubular flowers, somewhat four-angled stems, and somewhat clasping, simple, usually entire leaves; it is OBL. False Pimpernel (*Lindernia dubia*) has light purplish or whitish tubular flowers with five lobes forming two lips, upper lip two-lobed with a shallow notch and lower lip three-lobed; it is OBL.

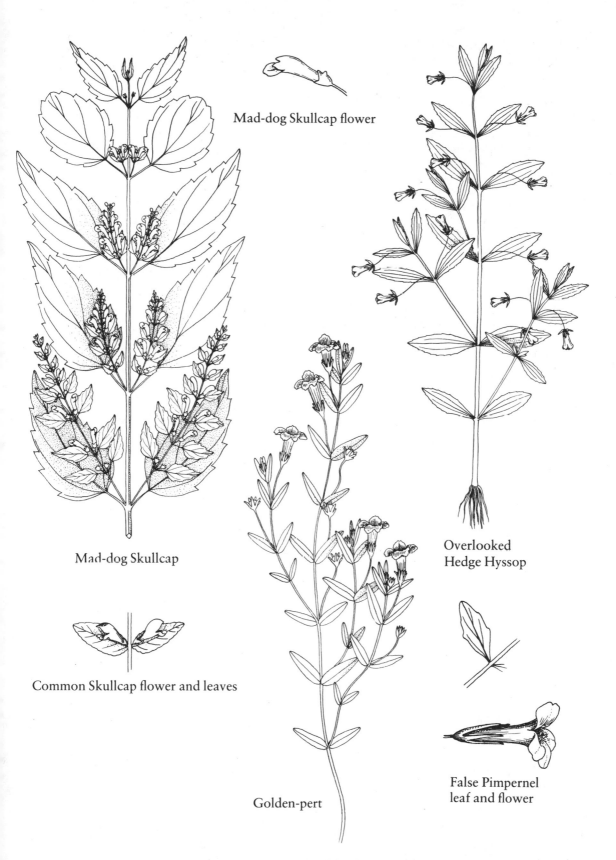

Mad-dog Skullcap flower

Mad-dog Skullcap

Common Skullcap flower and leaves

Golden-pert

Overlooked
Hedge Hyssop

False Pimpernel
leaf and flower

Flowering Herbs with Simple, Toothed, Opposite Leaves   237

False Nettle or Bog Hemp

Boehmeria cylindrica (L.) Swartz

Nettle Family
Urticaceae

Description: Medium-height, erect perennial herb, 1–3 feet tall; stems unbranched, smooth or rough-hairy; simple, coarse-toothed, somewhat broad lance-shaped leaves (1⅕–4⅘ inches long) tapering to a long point distally, petioled, with three distinct veins radiating from leaf base, oppositely arranged; inconspicuous greenish flowers borne in dense elongate spikes borne in leaf axils; shallow-winged oval nutlet (achene).

Flowering period: July into September.

Habitat: Tidal fresh marshes and swamps (regularly and irregularly flooded zones); nontidal marshes and swamps, hydric hammocks, and moist, usually shaded, soils.

Wetland indicator status: FACW+.

Range: Quebec and Ontario to Minnesota, south to Florida, Texas, and New Mexico.

Similar species: Clearweed (*Pilea pumila*) is similar but has translucent stems and glossy leaves; it is FACW.

Clearweed

Pilea pumila (L.) Gray

Nettle Family
Urticaceae

Description: Low to medium-height annual herb, up to 20 inches tall, often less than 12 inches; stems smooth, translucent; simple, coarse-toothed leaves (up to 5 inches long) with three main veins and long stalks, oppositely arranged; small greenish or whitish flowers borne in dense clusters at upper and middle leaf axils. (*Note:* Flower clusters are shorter than leaf stalks.)

Flowering period: July into September.

Habitat: Tidal swamps; nontidal forested wetlands along floodplains and cool, moist, shaded uplands.

Wetland indicator status: FACW.

Range: Quebec to Minnesota, south to Florida and Oklahoma.

Similar species: Other nettles lack translucent stems. Jewelweeds (*Impatiens* spp.) have somewhat translucent stems but bear distinctive, large orange or yellow tubular flowers.

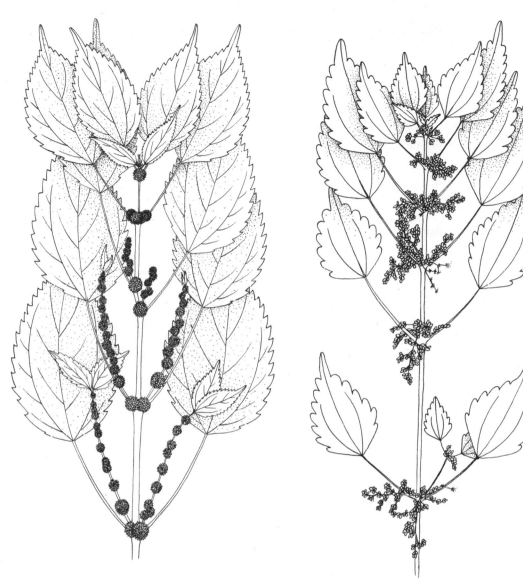

False Nettle

Clearweed

FLOWERING HERBS WITH COMPOUND ALTERNATE LEAVES

Sensitive Joint Vetch

Aeschynomene virginica (L.) B.S.P.

Legume Family
Leguminosae

Description: Medium-height to tall, erect annual herb, up to 5 feet high; stems branched and weakly bristle-hairy; pinnately compound leaves with odd number of numerous (up to fifty-six) oblong leaflets (½–1 inch long) with rounded tips, alternately arranged; one to six yellow or reddish flowers with two lips, short tube, and red veins, borne on inflorescences (racemes) in leaf axils; pealike fruit pod (1⅕–2½ inches long) with four to ten segments borne on a stalk (about ½–1 inch long).

Flowering period: July through October.

Habitat: Sandy or muddy tidal shores, tidal fresh marshes (regularly and irregularly flooded zones), and occasionally slightly brackish marshes.

Wetland indicator status: FACW.

Range: Southern New Jersey along the coast to North Carolina.

Similar species: Joint Vetch (*A. indica*) occurs in similar habitats from North Carolina to Florida and Texas; its leaflets (up to seventy in number) are mostly less than ½ inch long, and its fruit pod stalk is less than ½ inch long; it is FACW+. Partridge Pea (*Cassia fasciculata*) may also occur in tidal fresh marshes; it has five-petaled yellow flowers, and its pods are not distinctly segmented; it is FACU.

Water Hemlock or **Spotted Cowbane**

Cicuta maculata L.

Parsley Family
Umbelliferae

Description: Tall perennial herb, up to 6½ feet; stem smooth, hollow, branching, jointed near base, and purple-streaked; alternately arranged compound leaves divided into two or more coarse-toothed leaflets (up to 5 inches long); numerous small white flowers arranged in clusters (umbels, up to 5 inches wide); fruit oval, many-ribbed nutlet. (*Note:* Roots have a parsley aroma, and all plant parts are dangerously poisonous if eaten.)

Flowering period: May into September.

Habitat: Tidal fresh marshes; nontidal marshes, wet meadows, wet thickets, wooded swamps, and ditches.

Wetland indicator status: OBL.

Range: Quebec and Nova Scotia to Minnesota, south to Florida and Texas.

Similar species: Water Parsnip (*Sium suave*) has once-divided compound leaves with longer, narrower, more finely toothed leaflets; it is OBL.

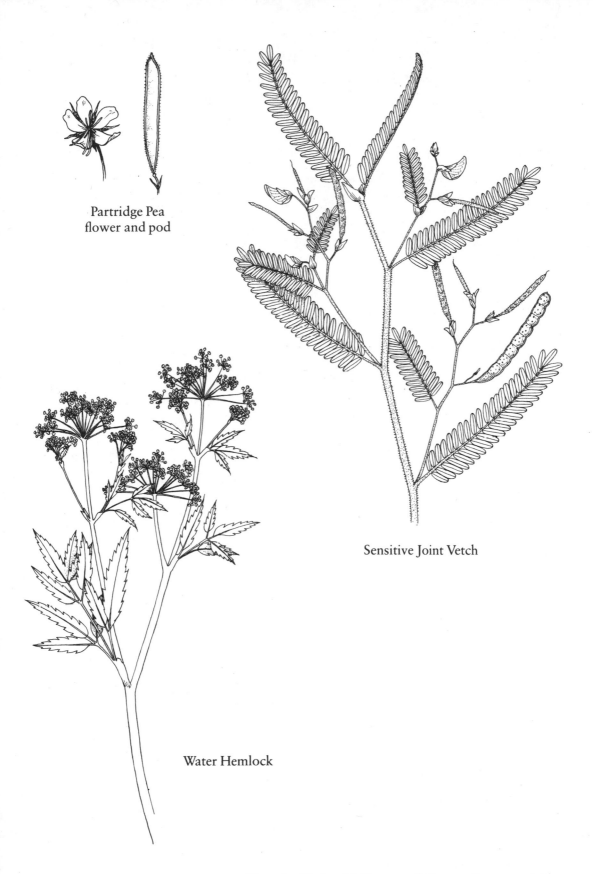

Partridge Pea
flower and pod

Sensitive Joint Vetch

Water Hemlock

Mock Bishopweed

Ptilimnium capillaceum (Michx.) Raf.

Parsley Family
Umbelliferae

Description: Low to medium-height, erect annual herb, 4–32 inches tall; compound leaves divided into threadlike leaflets (⅕–1 inch long), alternately arranged; very small white (rarely pinkish) flowers borne on umbels (⅘–2 inches wide) that overtop leaves; somewhat egg-shaped fruit with distinctive ribs and corky band.

Flowering period: May through October.

Habitat: Brackish and fresh tidal marshes (regularly and irregularly flooded zones); nontidal marshes.

Wetland indicator status: OBL.

Range: Massachusetts to Florida and Texas, chiefly along the coast; inland in southern states to Missouri and Oklahoma; also reported in South Dakota.

Water Parsnip

Sium suave Walter

Parsley Family
Umbelliferae

Description: Tall, erect perennial herb, up to 7 feet high; stems grooved or strongly angled and smooth; compound leaves with seven to seventeen leaflets (2–4 inches long), linear or lance-shaped, strongly toothed upper leaflets often simple, alternately arranged; very small white flowers borne in umbels (to about 5 inches wide); somewhat elongate oval fruit capsule with prominent ribs.

Flowering period: June through August.

Habitat: Slightly brackish marshes and tidal fresh marshes; nontidal marshes, swamps, and muddy shores.

Wetland indicator status: OBL.

Range: Newfoundland to British Columbia, south to Florida, Louisiana, and California.

Similar species: Water Hemlock (*Cicuta maculata*) occurs in freshwater marshes; its leaves may be once, twice, or thrice divided, some leaflets are three-lobed, and its stem is not strongly angled and may be purple-mottled; it is OBL. (*Caution:* Its fleshy roots are extremely poisonous.) Stiff Cowbane or Water Dropwort (*Oxypolis rigidior*) also has compound leaves with five to eleven leaflets, but its leaflet margins are either entire, mostly entire with a few scattered coarse teeth, or coarse-toothed; it is OBL. Water Cowbane (*O. filiformis*) occurs in brackish and tidal fresh marshes; it has thick, thread-like to linear, simple leaves with internal partitions (septate); it is FACW+.

Mock Bishopweed

Water Parsnip

Stiff Cowbane leaf

Devil's Beggar-ticks

Bidens frondosa L.

Composite or Aster Family
Compositae (Asteraceae)

Description: Medium-height to tall annual herb, up to 4 feet high; compound leaves divided into three to five coarse-toothed lance-shaped leaflets (up to 4 inches long and 1¼ inches wide) borne on stalks, oppositely arranged; yellow to somewhat orange disk flowers borne in dense heads (up to ½ inch wide) on leafy branches; barbed seeds.

Flowering period: June into October.

Habitat: Tidal fresh marshes; nontidal marshes, wet meadows, floodplain forests, ditches, fields, pastures, and waste places.

Wetland indicator status: FACW.

Range: Newfoundland and Nova Scotia to Washington, south to Georgia, Louisiana, and California.

Similar species: Flowers resemble Swamp Beggar-ticks (*B. connata*), but leaves of *B. connata* are simple and not compound; it is OBL. Nodding Beggar-ticks (*B. cernua*) has large yellow flowers with eight "petals" (ray flowers) surrounding head of disk flowers and simple leaves; it is OBL. Bearded Beggar-ticks (*B. aristosa*) is an annual with a taproot; it often has yellow "petals" (ray flowers), its leaves are divided into five to seven leaflets, and its nutlets are covered with fine hairs (cilia); it is FACW. Large-fruit Beggar-ticks (*B. coronota*) is similar to *B. aristosa*, except that its nutlets (longer than ⅕ inch) are not covered with cilia; it is OBL. Small-fruit Beggar-ticks (*B. mitis*) is like *B. coronota*, but its nutlets are smaller (⅕ inch or less long); it is OBL.

SEE ALSO Dog-fennel (*Eupatorium capillifolium*) described under Boneset (*E. perfoliatum*).

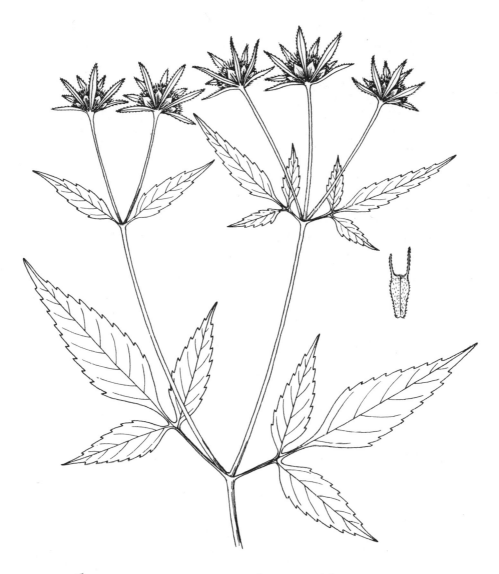

Devil's Beggar-ticks

flowers

THORNY SHRUBS

Swamp Rose

Rosa palustris Marshall
Rose Family
Rosaceae

Description: Broad-leaved, deciduous thorny shrub, up to 7 feet tall; stems much branched, bearing decurved thorns, upper branches smooth except for scattered thorns; compound leaves divided into seven finely toothed, dull green, narrowly egg-shaped leaflets; pink five-petaled flowers (1½–2½ inches wide) borne in small clusters or singly; hairy and red fleshy rose hip fruit enclosing numerous seeds.

Flowering period: May to October.

Fruiting period: Summer into winter.

Habitat: Tidal fresh marshes; nontidal forested wetlands, shrub swamps, marshes, and stream banks.

Wetland indicator status: OBL.

Range: Nova Scotia and New Brunswick to Minnesota, south to Florida and Arkansas.

Similar species: Multiflora Rose (*R. multiflora*), a native of eastern Asia and escapee from cultivation, may occur in tidal forested wetlands; it has fragrant white flowers (up to 1½ inches wide) and small berrylike rose hips (less than ⅓ inch wide). Gray Nicker or Nickerbean (*Caesalpina bonduc*) is a prickly vinelike tropical shrub found along the borders of mangrove swamps; its compound leaves are very long (to 2 feet) and divided into many pairs of leaflets, and its fruit pods are brown and densely prickled; it is FACU+.

SEE ALSO thorny trees: Devil's Walking Stick (*Aralia spinosa*) and Water Locust (*Gleditsia aquatica*).

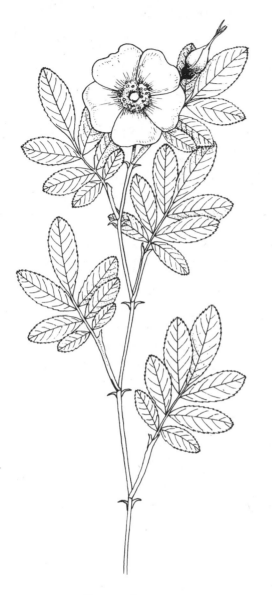

Multiflora Rose

Swamp Rose

Multiflora Rose hips

EVERGREEN SHRUBS

Inkberry

Ilex glabra (L.) Gray

Holly Family
Aquifoliaceae

Description: Broad-leaved evergreen shrub, up to 7 feet tall; simple, somewhat leathery, narrow leaves (up to 2 inches long) with one to three pairs of coarse teeth near tip, wedge-shaped bases, stalked, alternately arranged; small six- to eight-lobed white flowers borne singly or in clusters from leaf axils; black berries.

Flowering period: May into July.

Habitat: Tidal swamps; nontidal forested wetlands, shrub swamps, hydric hammocks, and sandy woods.

Wetland indicator status: FACW.

Range: Nova Scotia south along the Coastal Plain to Florida, west to Louisiana.

Similar species: Large or Sweet Gallberry (*I. coriacea*) has thick leathery evergreen few-toothed or entire leaves and bears black berries, but its leaves are large (to about 3½ inches long) and are dark green above with few short pricklelike marginal teeth mostly above the middle; its berries are sweet; it occurs along pond margins and in pine flatwoods and Carolina bays along the Coastal Plain from Virginia south; it is FACW.

Titi

Cyrilla racemiflora L.

Cyrilla Family
Cyrillaceae

Description: Broad-leaved semi-evergreen shrub or small tree, up to 27 feet tall; simple, entire, smooth, leathery egg-shaped to spoon-shaped leaves (up to 5 inches long and to 1 inch wide), with wedge-shaped or tapered bases, conspicuously netted-veined, alternately arranged; numerous minute five-petaled white flowers (less than ⅕ inch long) borne in many dense clusters on axillary inflorescences (racemes, up to 6 inches long) at end of previous year's twigs and below current year's leafy growth; brown or olive fruits (drupes, up to about ¹⁄₁₀ inch long) bearing one or two seeds.

Flowering period: May into July.

Habitat: Tidal swamps; pocosins, low pine flatwoods, and stream banks.

Wetland indicator status: FACW.

Range: Southeastern Virginia to central Florida, west to southeastern Texas; also in Mexico, West Indies, and South America.

SEE ALSO Wax Myrtle (*Myrica cerifera*), Sweet Bay (*Magnolia virginiana*), Brazilian Pepper (*Schinus terebinthifolius*) described under Black Mangrove (*Avicennia germinans*), and evergreen trees with needlelike or scalelike leaves.

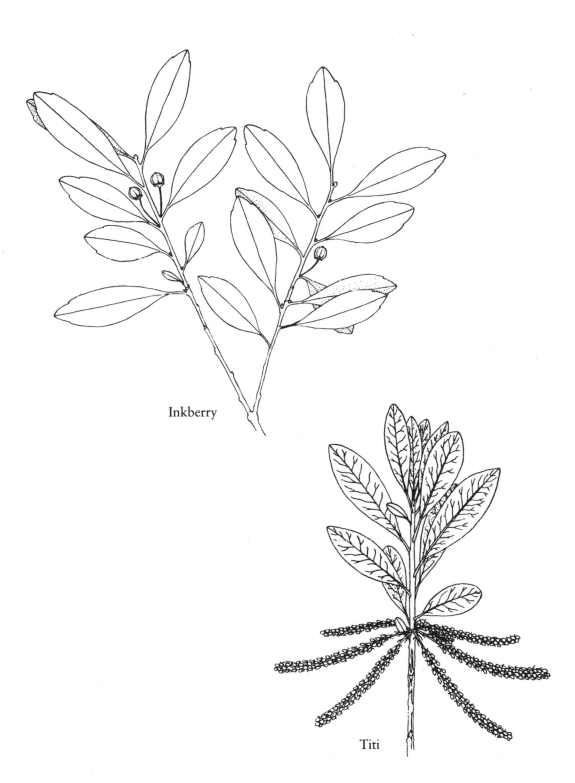

Inkberry

Titi

DECIDUOUS SHRUBS WITH COMPOUND LEAVES

Poison Sumac

Toxicodendron vernix (L.) Kuntze
[*Rhus vernix* L.]

Cashew Family
Anacardiaceae

Description: Broad-leaved deciduous shrub
or low tree, up to 23 feet tall; milky sap; alternately arranged compound leaves divided into
seven to thirteen entire, pointed leaflets (up to
3½ inches long); small greenish yellow flowers
in dense clusters (up to 8 inches long) borne on
long stalks from leaf axils; small whitish berries.
(*Warning: Do not touch*; plant is poisonous and
may cause severe skin irritations.)

Flowering period: May through June.

Habitat: Tidal swamps; nontidal seasonally
flooded forested wetlands and occasionally borders of marshes.

Wetland indicator status: OBL.

Range: Maine to Minnesota, south to Florida
and Texas.

Similar species: Poison Ivy (*T. radicans*) has
compound leaves divided into three leaflets; it
is FAC.

Common Elderberry

Sambucus canadensis L.

Honeysuckle Family
Caprifoliaceae

Description: Broad-leaved deciduous shrub,
up to 12 feet tall; multiple stems usually about
1 inch in diameter with thick, soft white center
(pith) and light brown bark with numerous large
raised bumps (lenticels); oppositely arranged,
compound leaves divided into five to eleven,
usually seven, fine-toothed, lance-shaped,
stalked leaflets (up to about 7 inches long) tapering to a distinct point distally, lower leaflets
sometimes divided into three parts; numerous
small white, five-lobed tubular flowers borne on
dense, somewhat flat-topped terminal inflorescences (cymes) with five spreading branches
from end of twig; fruit dark purplish berry.

Flowering period: May through August.

Fruiting period: Middle to late summer.

Habitat: Irregularly flooded tidal fresh marshes
and swamps; nontidal marshes, wet meadows,
swamps, hydric hammocks, old fields, moist
woods, and roadsides.

Wetland indicator status: FACW−.

Range: Nova Scotia to Manitoba and South Dakota, south to Florida and Texas.

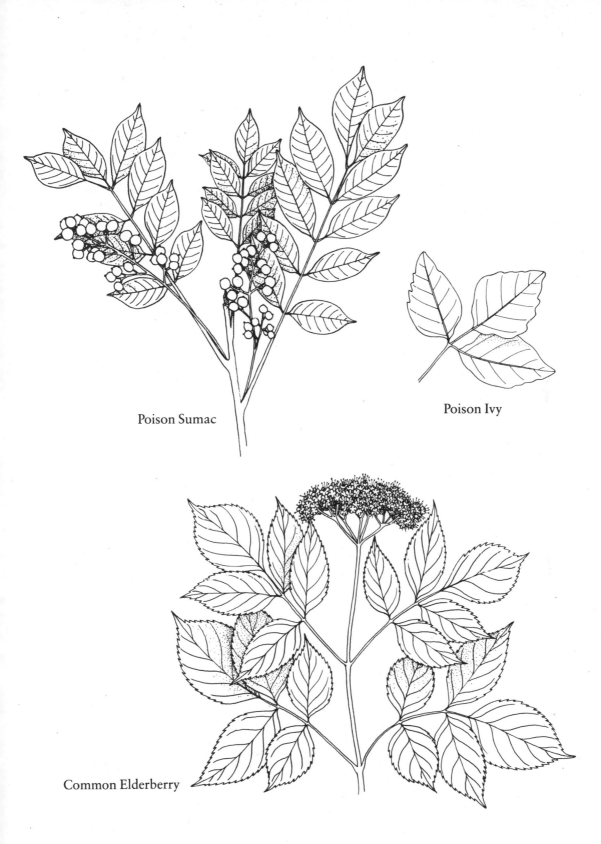

Poison Sumac

Poison Ivy

Common Elderberry

False Indigo

Amorpha fruticosa L.

Legume Family
Leguminosae

Description: Deciduous shrub, up to 16 feet tall; smooth dark gray bark; twigs round or finely grooved; alternately arranged compound leaves (up to 16 inches long) divided into eleven to thirty-five narrowly egg-shaped to oblong, short-stalked leaflets (up to 1⅓ inches long and to ½ inch wide), somewhat pointed or blunt distally, margins entire, dull green above and usually weakly hairy below; numerous small purplish flowers borne on dense, erect, spikelike inflorescences (racemes, up to 8 inches long); small olive fruit pods marked with red dots.

Flowering period: April into August.

Habitat: Irregularly flooded tidal fresh marshes and swamps; inland moist woods, riverbanks, and lake shores.

Wetland indicator status: FACU.

Range: New England to Minnesota and Saskatchewan, south to Florida and Texas.

False Indigo

DECIDUOUS SHRUBS WITH SIMPLE, ENTIRE, ALTERNATE LEAVES

Swamp Azalea

Rhododendron viscosum (L.) Torr.

Heath Family
Ericaceae

Description: Broad-leaved deciduous shrub, up to 10 feet tall; simple, entire, shiny green leaves (up to 2½ inches long), with hairy margins and a lower midrib, alternately arranged; tubular five-lobed, fragrant, sticky white flowers (about 1 inch wide and 1 inch long) borne in clusters at ends of branches; persistent five-parted fruit capsules.

Flowering period: May through October (after leaf-out).

Habitat: Tidal swamps; nontidal forested wetlands and occasionally sandy upland woods.

Wetland indicator status: FACW+.

Range: Maine to Ohio, south to Florida.

Similar species: Hoary Azalea (*R. canescens*) bears pink and white flowers before leaves emerge; it is common in forested wetlands along the Coastal Plain; it is FACW−.

Swamp Azalea

Highbush Blueberry

Vaccinium corymbosum L.

Heath Family
Ericaceae

Description: Broad-leaved deciduous shrub, up to 13 feet tall, often growing in clumps; simple, entire or sharp-toothed leaves (up to 3¼ inches long), paler below, alternately arranged; small whitish (sometimes pink-tinged) urn-shaped flowers (up to ½ inch long) borne in dense terminal and lateral clusters; blue berries (up to ½ inch wide).

Flowering period: February through May. (*Note:* Flowers appear as leaves are emerging.)

Habitat: Tidal swamps; nontidal forested wetlands, shrub swamps, bogs, and upland woods (uncommon).

Wetland indicator status: FACW.

Range: Nova Scotia to Wisconsin, south to Florida and Texas.

Similar species: Huckleberries (*Gaylussacia* spp.) are similar, but they have yellowish resin dots on their leaves.

Corkwood

Leitneria floridana Chapm.

Corkwood Family
Leitneriaceae

Description: Broad-leaved deciduous shrub, up to 20 feet tall or more; young bark smooth, brownish with conspicuous light-colored vertical lenticels; simple, entire, thick, lance-shaped to narrowly egg-shaped leaves (up to 8 inches long and to 2 inches wide) smooth above and somewhat soft-hairy below (at least on veins), alternately arranged; inconspicuous flowers borne on elongate cylinder-shaped spikes (catkins, ½–2 inches long), male and female flowers typically borne on separate plants (dioecious); somewhat oblong brownish berry (up to 1 inch long).

Flowering period: Spring.

Habitat: Brackish and tidal fresh marshes and swamps; sawgrass marshes, nontidal swamps, and wet prairies.

Wetland indicator status: OBL.

Range: Southern Georgia to northern Florida, west to Texas and southeastern Missouri.

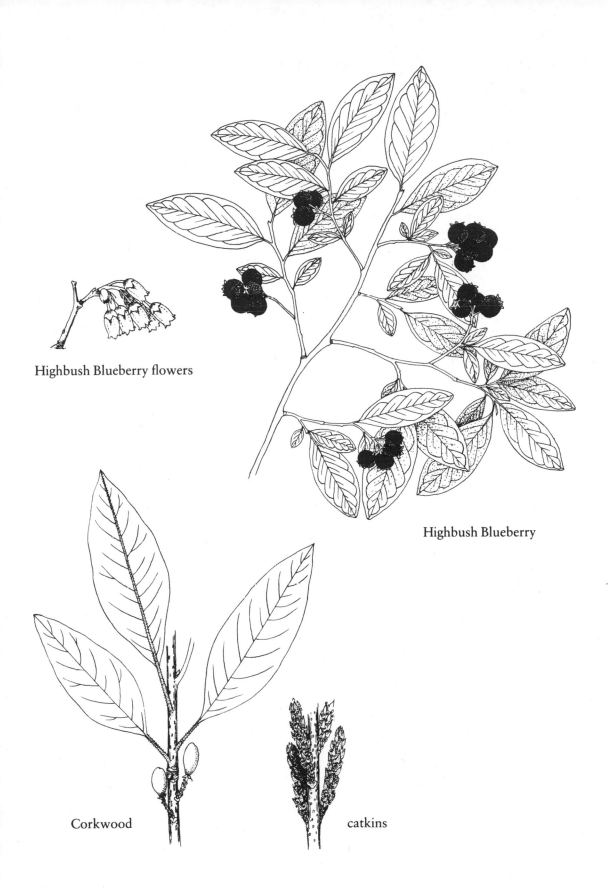

Highbush Blueberry flowers

Highbush Blueberry

Corkwood

catkins

DECIDUOUS SHRUBS WITH SIMPLE, ENTIRE, OPPOSITE LEAVES

Southern Wild Raisin or Possum-haw

Viburnum nudum L.

Honeysuckle Family
Caprifoliaceae

Description: Broad-leaved deciduous shrub, up to about 16 feet tall; simple, mostly entire to wavy-margined (sometimes toothed), leathery, somewhat egg-shaped leaves (up to 6 inches long and to 2½ inches wide), widest at middle, shiny green above, usually dotted below, with pointed tips and usually wedge-shaped bases, stalked (winged stalks up to 1 inch long), oppositely arranged; small white (rarely pink) five-"petaled" flowers borne in flat-topped clusters (cymes, 2–6 inches wide) from leaf axils; bluish black berries (about ⅓ inch long) covered with waxy bloom and bearing one seed.

Flowering period: March to June.

Habitat: Tidal fresh marshes and swamps; nontidal swamps, pocosins, wet pine flatwoods, and low woods.

Wetland indicator status: FACW+.

Range: Long Island, New York, south along the Coastal Plain to Florida and eastern Texas.

Similar species: Swamp Dogwood (*Cornus foemina*, formerly *C. stricta*) has opposite entire leaves with prominent long tips, but its leaves are not thick leathery or shiny; it bears clusters of small four-petaled white flowers at the end of young branches; if a leaf is gently pulled apart from the middle, strands of the lateral veins usually remain connected; it is a common shrub in swamps and on river banks along the Coastal Plain from Delaware south; it is FACW−.

Buttonbush

Cephalanthus occidentalis L.

Madder Family
Rubiaceae

Description: Broad-leaved deciduous shrub, 3½–10 feet tall; young bark smooth and grayish, older bark grayish brown and flaky; pith light brown; twigs grayish brown to purplish, round, hairy or smooth, and marked with light elongate dots (lenticels); simple, entire, egg-shaped leaves (3–6 inches long) tapering to a short point, oppositely arranged but sometimes in whorls of threes and fours, leaf stalks often red; small white tubular flowers in dense ball-shaped heads (about 1 inch wide); nutlet-bearing fruit ball.

Flowering period: June through August.

Habitat: Tidal fresh marshes; nontidal marshes, shrub swamps, forested wetlands, wet prairies, and borders of streams, lakes, and ponds.

Wetland indicator status: OBL.

Range: New Brunswick and Quebec to Minnesota, south to Florida, Mexico, and California.

SEE ALSO Sandweed St. John's-wort (*Hypericum fasciculatum*) described under Dwarf St. John's-wort (*H. mutilum*).

Southern Wild Raisin

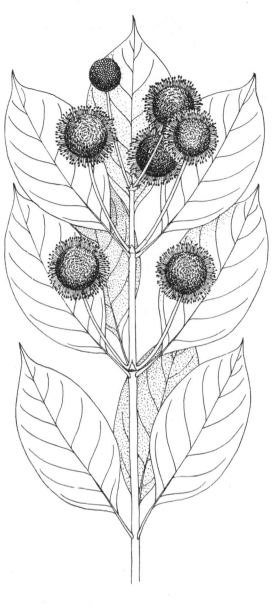

Swamp Dogwood

Buttonbush

DECIDUOUS SHRUBS WITH SIMPLE, TOOTHED, ALTERNATE LEAVES

Common Winterberry

Ilex verticillata (L.) Gray

Holly Family
Aquifoliaceae

Description: Deciduous shrub, up to 16 feet tall; bark dark gray and smooth; twigs gray and smooth; simple, coarse-toothed, egg-shaped to oblong lance-shaped leaves (2–4 inches long and about ½ inch wide) tapering distally to a prominent short point, dull green above, wrinkled below, somewhat wedge-shaped base, petioled, alternately arranged; two types of flowers (male and female) borne on separate plants (dioecious), small white flowers with four to six petals on short stalks borne singly or in clusters (female flowers usually solitary and male flowers in clusters), petals slightly joined at base; fruit bright red, rarely yellow, berrylike (drupe).

Flowering period: April to June.

Range: Irregularly flooded tidal swamps and upper borders of tidal fresh marshes; nontidal shrub swamps and forested wetlands.

Wetland indicator status: FACW.

Range: Newfoundland to Minnesota, south to Georgia and Mississippi.

Similar species: Inkberry (*I. glabra*) has smooth, shiny, leathery evergreen leaves with one to three distal teeth on each side margin, small white flowers, and black berrylike fruits; it is FACW.

Smooth Alder

Alnus serrulata (Ait.) Willd.

Birch Family
Betulaceae

Description: Tall deciduous shrub, up to 20 feet or more in height; multiple trunks, dark gray bark marked with small lighter dots (lenticels); twigs dark grayish brown with lenticels and smooth; simple, fine-toothed, egg-shaped leaves (up to 5⅕ inches long and to 3⅕ inches wide), smooth above and usually hairy along veins below, stalked, alternately arranged; minute flowers borne in two types of dense spikes (catkins), male catkins elongate (less than 1 inch long), female catkins oval-shaped and appearing conelike after releasing seeds, both spikes persisting through winter; winged nutlet.

Flowering period: February to April.

Habitat: Irregularly flooded tidal fresh marshes and swamps; nontidal marshes, swamps, and stream banks.

Wetland indicator status: FACW+.

Range: Maine to New York and Missouri, south to Florida, Texas, and Oklahoma.

Common Winterberry

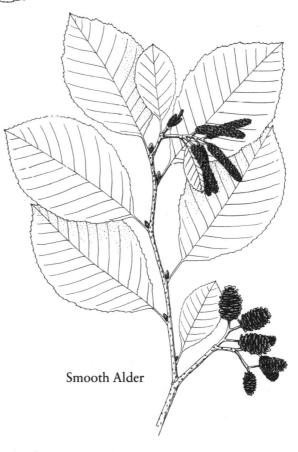

Smooth Alder

Sweet Pepperbush

Clethra alnifolia L.

White Alder Family
Clethraceae

Description: Broad-leaved deciduous shrub, up to about 10 feet tall; bark grayish brown and flaky; simple, coarse-toothed, somewhat egg-shaped to oblong leaves (to 4¾ inches long and to 1¾ inches wide) tapering to a fine point distally and wedge-shaped near base, widest above middle, toothed along upper leaf margin, lower margin mostly entire, usually short-stalked, alternately arranged; numerous small five-petaled, fragrant white flowers borne on terminal spike-like, short-hairy inflorescences (racemes, up to 8 inches long); hairy three-valved capsule (persists through winter).

Flowering period: July through September.

Habitat: Irregularly flooded tidal swamps and upper edges or higher elevations within tidal fresh marshes; nontidal forested wetlands, shrub swamps, pocosins, Carolina bays, wet pinelands, and sandy woods.

Wetland indicator status: FACW.

Range: Southern Maine to northern Florida and Texas.

Fetterbush or Swamp Sweetbells

Leucothoe racemosa (L.) Gray

Heath Family
Ericaceae

Description: Broad-leaved deciduous shrub, up to 13 feet tall; simple, obscurely toothed leaves (up to 3¼ inches long), short-stalked, alternately arranged; small white (sometimes pink-tinged), urn-shaped, five-lobed flowers (⅜–½ inch long) borne on dense one-sided clusters from leaf axils; five-valved fruit capsules (³⁄₁₆ inch wide).

Flowering period: March through June. (*Note:* Flower buds develop during previous summer.)

Habitat: Tidal swamps; nontidal shrub swamps, forested wetlands, and moist acid woods.

Wetland indicator status: FACW.

Range: Eastern Massachusetts south to Florida and Louisiana (mostly along the Coastal Plain).

Similar species: Maleberry (*Lyonia ligustrina*) has small white globe-shaped flowers borne in dense branched clusters; it is FACW.

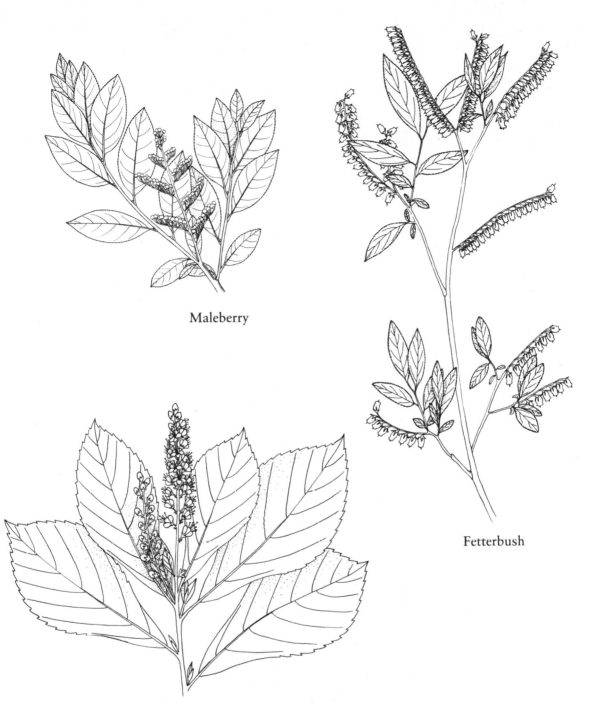

Maleberry

Fetterbush

Sweet Pepperbush

Oblong-leaf Juneberry
or Shadbush

Amelanchier canadensis (L.) Medic.

Rose Family
Rosaceae

Description: Broad-leaved deciduous shrub or small tree, up to 25 feet tall, usually growing in clumps; simple, fine-toothed oblong leaves (up to 3¼ inches long) rounded at tip and base, stalked, alternately arranged; numerous medium-sized five-petaled white flowers (¾ inch wide) on stalks borne in clusters; dark purple to black berries.

Flowering period: March through April.

Habitat: Tidal swamps; nontidal shrub swamps, forested wetlands, and upland woods.

Wetland indicator status: FAC.

Range: Newfoundland south along the Coastal Plain to Mississippi.

Virginia Sweet-spires
or Virginia Willow

Itea virginica L.

Saxifrage Family
Saxifragaceae

Description: Broad-leaved deciduous shrub, up to 10 feet tall; hairy twigs and inflorescences and white chambered pith; simple, fine-toothed elliptic leaves (up to 4 inches long), stalked, alternately arranged; small five-petaled white flowers borne in many-flowered spikelike inflorescences (up to 8 inches long); elongate fruit capsules.

Flowering period: May through June.

Habitat: Tidal swamps; nontidal forested wetlands and hydric hammocks.

Wetland indicator status: FACW+.

Range: Southern New Jersey south along Coastal Plain to Florida and Louisiana, north in Mississippi Valley to Illinois.

SEE ALSO Highbush Blueberry (*Vaccinium corymbosum*), Black Willow (*Salix nigra*), Swamp Willow (*S. caroliniana*), described under Black Willow, and Water Elm (*Planera aquatica*) described under American Elm (*Ulmus americana*).

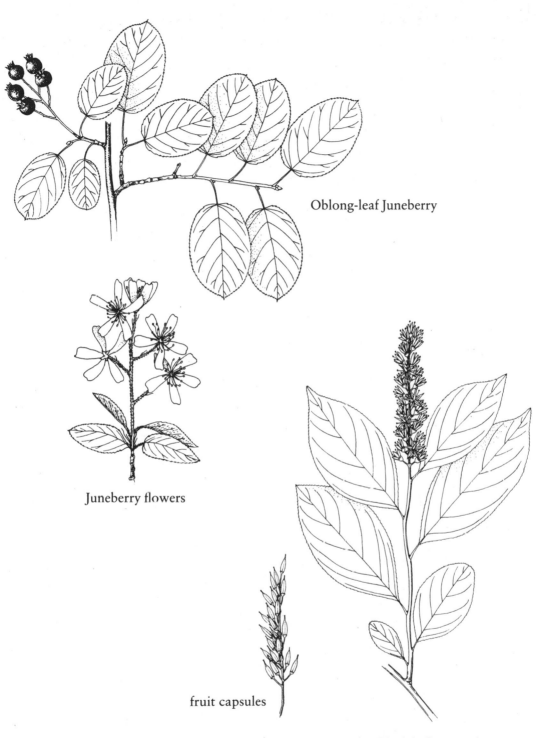

Oblong-leaf Juneberry

Juneberry flowers

fruit capsules

Virginia Sweet-spires

DECIDUOUS SHRUBS WITH SIMPLE, TOOTHED, OPPOSITE LEAVES

Southern Arrowwood

Viburnum dentatum L.

Honeysuckle Family
Caprifoliaceae

Description: Broad-leaved deciduous shrub, up to 12 feet tall; hairy twigs, simple, coarse-toothed, somewhat egg-shaped to round leaves (up to 4 inches long), hairy beneath, oppositely arranged; small five- to seven-petaled white flowers (¼ inch wide) in dense branched clusters borne on long stalks at end of branches; bluish black berries.

Flowering period: March into August.

Habitat: Tidal swamps; nontidal forested wetlands, shrub swamps, dry woods, and sandy soils, especially along Coastal Plain.

Wetland indicator status: FAC.

Range: Massachusetts south to Florida and Texas.

Similar species: See Southern Wild Raisin (*V. nudum*).

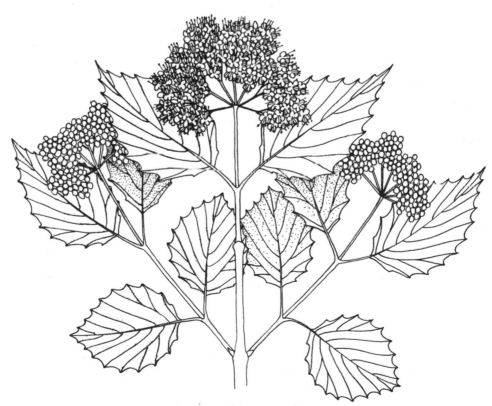

Southern Arrowwood

THORNY TREES

Devil's Walking Stick

Aralia spinosa L.

Ginseng Family
Araliaceae

Description: Broad-leaved deciduous thorny
shrub or tree, up to 40 feet tall; stout thorns on
stem, branches, and leaf stalks; large, alternately
arranged, compound leaves divided into many
coarse-toothed leaflets (up to 3½ inches long);
many small white flowers borne in many compound umbels forming a terminal inflorescence;
black berries.

Flowering period: June through September.

Habitat: Tidal swamps; nontidal forested wetlands and moist upland woods.

Wetland indicator status: FAC.

Range: Southern New England to Missouri,
south to Florida and Texas.

Similar species: Hercules' Club *Zanthoxylum
clava-herculis*) occurs in sand dunes and maritime forests; its leaf stalk is thorny, and its leathery compound leaves are round-toothed; it is
usually a shrub or short tree; it is FAC.

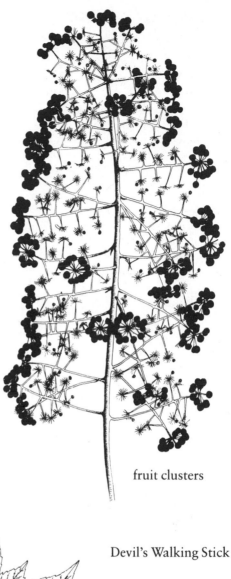

fruit clusters

Devil's Walking Stick

Water Locust

Gleditsia aquatica Marshall

Legume Family
Leguminosae

Description: Broad-leaved deciduous tree, up to 80 feet tall; bark gray, brown, or blackish, smooth, warty, or furrowed; stout branched or unbranched spines (up to 5½ inches long) on bark and branches; compound leaves (to 8 inches long) divided into many (twelve to eighteen) weakly round-toothed to wavy-margined, lance-shaped to oblong leaflets (to about 1½ inches long and to about ½ inch wide), smooth lower midvein, alternately arranged; few to many yellowish to greenish yellow, three- to five-petaled bell-shaped flowers (about ¼ inch wide) borne in clusters (racemes) in leaf axils; thin, flattened, somewhat oval-shaped fruit pod (1–2 inches long and to about 1½ inches wide) bearing one to three seeds (lacking pulp covering).

Flowering period: April and May.

Habitat: Tidal fresh swamps; nontidal river swamps, floodplain forests, and hydric hammocks, often in areas subject to long periods of inundation.

Wetland indicator status: OBL.

Range: South Carolina to central Florida, west to eastern Texas and north along Mississippi Valley to southern Illinois and Indiana.

Similar species: Honey Locust (*G. triacanthos*) has a hairy lower midvein, more leaflets (eighteen to twenty-eight), and an elongated fruit pod (up to 16 inches long and about 1½ inches wide) bearing many pulp-covered seeds; it is FAC−.

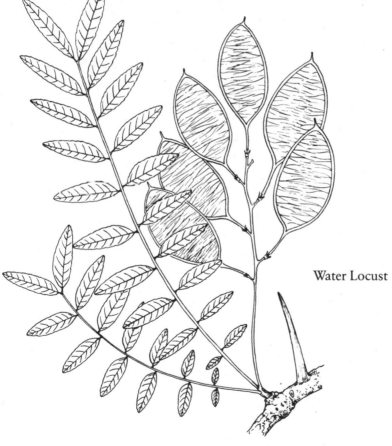

Water Locust

EVERGREEN TREES WITH NEEDLELIKE OR SCALELIKE LEAVES

Atlantic White Cedar

Chamaecyparis thyoides (L.) B.S.P.

Pine Family
Pinaceae

Description: Scale-leaved evergreen coniferous tree, up to 90 feet tall; reddish brown shaggy bark; flattened or four-angled twigs; scalelike evergreen leaves (to ⅛ inch long), bluish green to pale green, aromatic, appressed and completely covering twigs, oppositely arranged; small globe-shaped cones (¼–⅜ inch wide) with short-pointed scales.

Flowering period: March through April.

Fruiting period: April into fall.

Habitat: Tidal swamps; nontidal seasonally flooded forested wetlands, shrub bogs, and edges of streams (mostly on the Coastal Plain).

Wetland indicator status: OBL.

Range: Central Maine south mostly along the Coastal Plain to Florida and Mississippi.

Similar species: See Eastern Red Cedar (*Juniperus virginiana*).

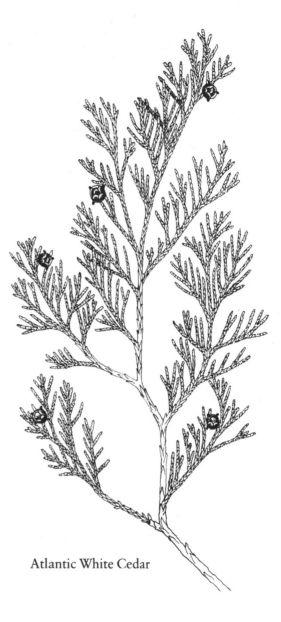

Atlantic White Cedar

Loblolly Pine

Pinus taeda L.

Pine Family
Pinaceae

Description: Needle-leaved evergreen coniferous tree, up to 115 feet tall; reddish brown to blackish gray, deeply furrowed bark forming large plates; long, stiff, often twisted yellowish green needles (up to 10 inches long) in bundles of three (rarely in twos); large spiny cone (up to 5 inches long).

Flowering period: March and April.

Habitat: Tidal swamps; nontidal forested wetlands, hydric hammocks, moist sandy soil of the Coastal Plain, abandoned fields, and pine plantations.

Wetland indicator status: FAC.

Range: Southern New Jersey to Florida and Texas, inland to Tennessee and Oklahoma.

Similar species: Pond Pine (*P. serotina*) has smaller cones (to 2½ inches long) that have few to no spines; it is OBL. Slash Pine (*P. elliottii*) has shiny dark green needles in bundles of three (sometimes in twos), shiny reddish brown spiny cones, and purplish brown peeling bark; it is FACW. Australian Pine (*Casuarina equisetifolia*), a pinelike exotic tropical tree, has colonized higher elevations in and adjacent to mangrove swamps; its evergreen needles are jointed; it is FACU.

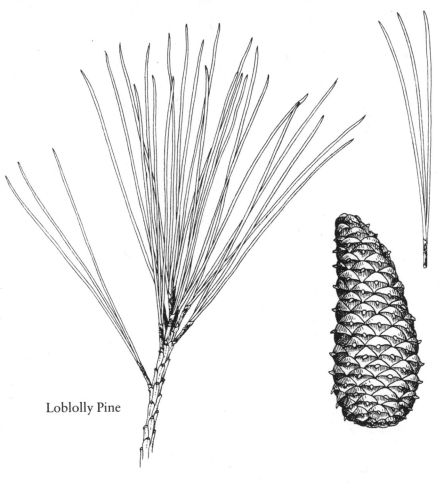

Loblolly Pine

EVERGREEN TREES WITH BROAD LEAVES

Sweet Bay or Swamp Magnolia

Magnolia virginiana L.

Magnolia Family
Magnoliaceae

Description: Broad-leaved deciduous (north) or evergreen (south) shrub or tree, up to 70 feet tall; bark smooth and gray; twigs smooth, dark green, and aromatic (when crushed); chambered pith; simple, entire, leathery oblong leaves (up to 7 inches long), shiny green above and whitish hairy below, mildly aromatic when crushed, alternately arranged; large fragrant nine- to twelve-petaled white flowers (2–2½ inches wide); pink or red conelike fruit clusters.

Flowering period: April through July.

Habitat: Tidal swamps; nontidal forested wetlands, hydric hammocks, stream and pond borders, and moist sandy woods.

Wetland indicator status: FACW+.

Range: Long Island, New York, south to Florida and west to Texas (chiefly along the Coastal Plain).

Similar species: Red Bay or Swamp Bay (*Persea borbonia*, formerly *P. palustris*) has thick, leathery, strongly aromatic leaves with wedge-shaped bases, rusty-haired midvein below, and often densely hairy stalks, densely hairy young twigs, small flowers borne in clusters on hairy stalks, and dark blue to black roundish berries (about ⅓ inch wide); it is FACW. Loblolly Bay (*Gordonia lasianthus*) has nonaromatic, thick, evergreen leaves with round-toothed margins and its showy white flowers (2½–3 inches wide) are five-petaled; it is FACW. Southern Magnolia or Bull Bay (*M. grandifolia*) occurs in maritime forests and hydric hammocks and along the edges of southern swamps; its thick leathery leaves are glossy dark green above and rusty-hairy below; its fragrant white flowers are much larger than those of *M. virginiana*; it is FAC+.

fruit cluster

Red Bay

Sweet Bay

SEE ALSO Wax Myrtle (*Myrica cerifera*) and Titi (*Cyrilla racemiflora*).

Bald Cypress

Taxodium distichum (L.) Rich.

Redwood Family
Taxodiaceae

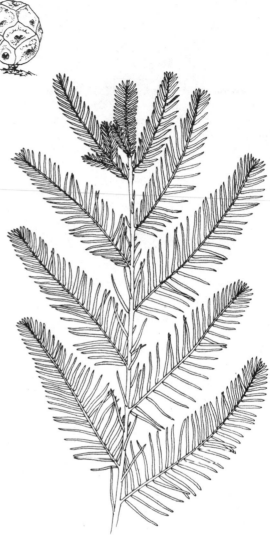

Description: Needle-leaved deciduous coniferous tree, up to 140 feet tall; cypress "knees" arising from roots around base of tree; reddish brown to gray scaly ridged bark, often shredding; usually buttressed trunks; simple, flattened linear, needlelike leaves (⅜–⅞ inch long) borne in two ranks, making twigs appear featherlike, yellowish to pale green above and whitish below, alternately arranged; inconspicuous flowers borne on cones; round cones (to 1 inch wide).

Flowering period: March and April.

Habitat: Tidal swamps; nontidal forested wetlands (especially seasonally flooded wetlands), riverbanks, and sometimes in permanent open water.

Wetland indicator status: OBL.

Range: Southern Delaware south (mostly along the Coastal Plain) to Florida and Texas, and in the Mississippi Valley north to southern Illinois and Indiana.

Similar species: Variety *nutans* called Pond Cypress has mostly appressed (not two-ranked) shorter needles (⅜ inch or less wide) and ascending branchlets; it occurs from southeastern Virginia south to Florida and Louisiana; it is OBL.

Bald Cypress

DECIDUOUS TREES WITH COMPOUND OPPOSITE LEAVES

Green Ash

Fraxinus pennsylvanica Marshall var.
  subintegerrima (Vahl) Fernald

Olive Family
Oleaceae

Description: Broad-leaved deciduous tree, up
to 80 feet tall; bark brown, shallowly grooved;
twigs gray and smooth; oppositely arranged,
compound leaves divided into five to nine, usually seven, shallow-toothed lance-shaped leaflets
(up to 6 inches long) tapering to a blunt or fine
point distally; flowers inconspicuous; winged
fruits (samaras).

Flowering period: April and May. (*Note:* Flowers appear as leaves emerge.)

Fruiting period: Summer to fall.

Habitat: Tidal swamps and higher areas within
and borders of irregularly flooded tidal fresh
marshes; nontidal swamps and hydric hammocks.

Wetland indicator status: FACW.

Range: Maine to Ontario and Saskatchewan,
south to Florida and Texas.

Similar species: Pumpkin Ash (*F. profunda*, formerly *F. tomentosa*) also has densely hairy
leaves and twigs, but its leaves are thick, leathery, and longer (10–20 inches), and its trunks
are buttressed (swollen); it is common in cypress–tupelo swamps; it is OBL. Carolina Ash
(*F. caroliniana*) has leathery leaves (7–12 inches
long) with coarse-toothed margins, but its two-
to three-winged fruits are broader (¾ inch wide)
than the other ashes (⅜ inch or less); it is OBL.
Water Hickory (*Carya aquatica*) is another com-
pound-leaved tree of tidal swamps; its leaves (9–
15 inches long) are alternately arranged and
composed of seven to fifteen (usually eleven to
thirteen) sickle-shaped leaflets, and its bark is
very shaggy; it is OBL.

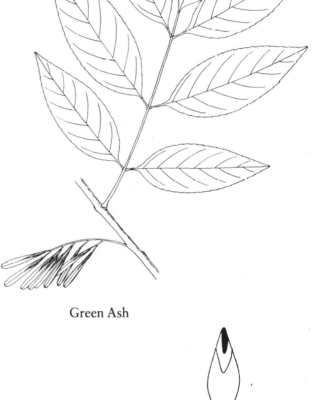

Green Ash

Carolina Ash samara

DECIDUOUS TREES WITH SIMPLE OPPOSITE LEAVES

Red Maple

Acer rubrum L.

Maple Family
Aceraceae

Description: Broad-leaved deciduous
shrub or tree, up to 120 feet tall;
smooth gray bark when young, broken and darker when older; young
twigs reddish, often partly covered
with whitish flaky coating; buds reddish, often clustered near tip of twigs;
simple coarse-toothed leaves, usually
with three to five shallow lobes (2–8
inches long and to about 7 inches
wide), oppositely arranged; small red
flowers in short clusters; fruits reddish and winged (samaras). (*Note:*
Variety *trilobum* has leaves with
three lobes; variety *drummondii* has
leaves with hairy undersides.)

Flowering period: January through May. (*Note:*
Red flowers appear before leaf buds open.)

Habitat: Tidal fresh marshes and swamps; nontidal swamps, alluvial soils, hydric hammocks,
stream banks, stable dunes, and moist uplands.

Wetland indicator status: FAC; varieties *drummondii* and *trilobum* are OBL.

Range: Newfoundland and Quebec to Manitoba and Minnesota, south to Florida and
Texas.

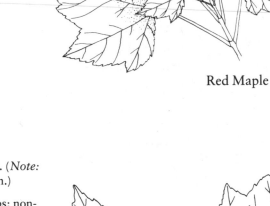

Red Maple

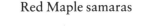

Red Maple samaras

Red Maple
variety trilobum

DECIDUOUS TREES WITH SIMPLE, ENTIRE, ALTERNATE LEAVES

Black Gum

Nyssa sylvatica Marshall

Dogwood Family
Cornaceae

Description: Broad-leaved deciduous tree, up to 125 feet tall; bark dark brown or gray, deeply furrowed; chambered pith; simple, entire (rarely few-toothed), somewhat leathery leaves (up to 6 inches long), shiny green above, alternately arranged; small greenish flowers (male and female flowers usually on different trees), many male flowers borne on stalks; bluish black berries.

Flowering period: April through June.

Habitat: Tidal swamps; nontidal forested wetlands, hydric hammocks, moist upland woods, and dry woods.

Wetland indicator status: FAC.

Range: Maine to southern Ontario, south to Florida and Texas.

Similar species: Variety *biflora* called Swamp Tupelo is the common wetland variety from Delaware south; its leaves are somewhat narrower and not short-pointed, and its fruiting stalks are less than 1⅕ inches long; it is OBL. Water Gum or Tupelo (*N. aquatica*) has larger leaves with few teeth, blue or purple berries, tapering leaf tips, and smooth or sparsely haired leaf stalks; it occurs in Coastal Plain swamps along rivers from southeastern Virginia south; it is OBL. Ogeechee Tupelo (*N. ogeche*) is similar to *N. aquatica*; it has red berries, round leaf tips with an abrupt point (mucronate tip), and ash-colored stalks; it is OBL.

Black Gum

Black Gum flowers

Water Gum

Persimmon

Diospyros virginiana L.

Ebony Family
Ebenaceae

Description: Broad-leaved deciduous tree, up to 70 feet tall; dark brown to black bark divided into squarish plates; pith solid, sometimes weakly chambered; simple, entire, somewhat leathery egg-shaped leaves (up to 6 inches long), long-pointed, shiny dark green above, alternately arranged; fragrant four-lobed, bell-shaped, greenish yellow flowers borne singly or in clusters; round orange fleshy fruits.

Flowering period: May through June.

Habitat: Tidal swamps; nontidal forested wetlands (often adjacent to salt marshes), hydric hammocks, moist alluvial woods, and dry uplands.

Wetland indicator status: FAC.

Range: Southern Connecticut to Iowa and Kansas, south to Florida and Texas.

Similar species: Resembles Black Gum (*Nyssa sylvatica*), but Black Gum's pith is chambered and its fruits are bluish berries; it is FAC.

Willow Oak

Quercus phellos L.

Beech Family
Fagaceae

Description: Broad-leaved deciduous tree up to 100 feet tall; dark gray to blackish, ridged bark; simple, entire or weakly wavy-edged, linear to linear lance-shaped leaves (up to 5 inches long and to 1 inch wide) bristle-tipped, alternately arranged; inconspicuous flowers borne on catkins (male) and singly or in clusters (female); acorns (⅜–½ inch long).

Flowering period: March into May.

Habitat: Tidal swamps (at higher elevations); forested wetlands (mostly temporarily flooded) along floodplains, and moist alluvial woods.

Wetland indicator status: FACW−.

Range: Southern New York to southern Illinois, south to Florida and Texas.

Similar species: Laurel Oak (*Q. laurifolia*) has narrowly egg-shaped (sometimes rather elongate diamond-shaped) leaves with wedge-shaped bases that are deciduous, yet usually persist through winter into early spring; common tree of southern swamps that has been reported along the edges of mangrove swamps and in hydric hammocks; it is FACW. Water Oak (*Q. nigra*) occurs in hydric hammocks near salt marshes and may be found in temporarily flooded tidal swamps; its leaves are widest near the top and have wedge-shaped bases and bristle tips; it is FAC.

SEE ALSO Sweet Bay (*Magnolia virginiana*).

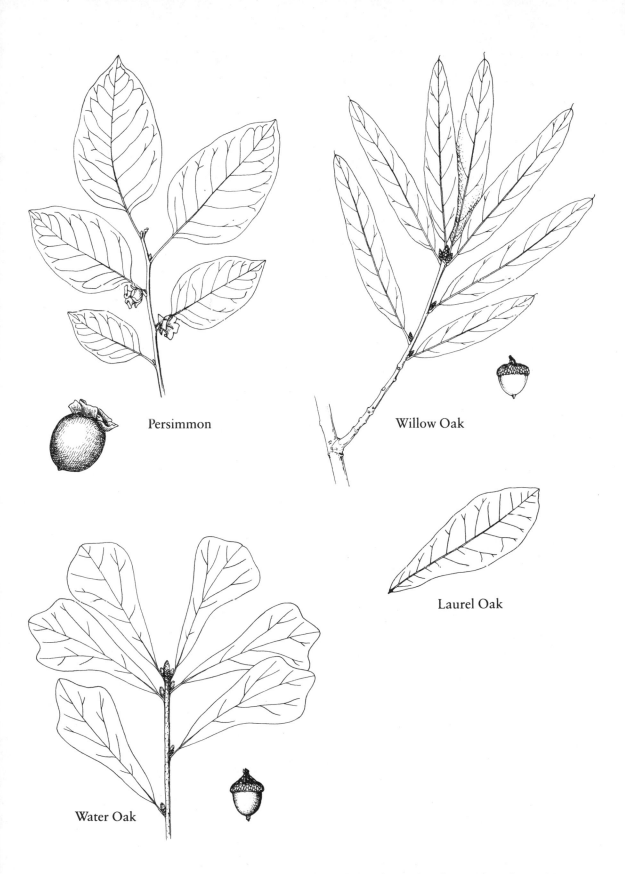

Persimmon

Willow Oak

Laurel Oak

Water Oak

DECIDUOUS TREES WITH SIMPLE, TOOTHED, UNLOBED ALTERNATE LEAVES

River Birch

Betula nigra L.

Birch Family
Betulaceae

Description: Broad-leaved deciduous tree, up to 100 feet tall; reddish brown, greatly peeling young bark, older bark dark black and plate-like; simple, irregularly coarse-toothed, triangle-shaped leaves (up to 5 inches long), whitish (hairy when young) below, alternately arranged; inconspicuous flowers borne on catkins; flattened winged fruits.

Flowering period: March through April.

Habitat: Tidal swamps; nontidal forested wetlands and floodplain forests.

Wetland indicator status: FACW.

Range: New Hampshire to Minnesota, south to Florida and Texas.

Ironwood, Musclewood, or Blue Beech

Carpinus caroliniana Walter

Birch Family
Betulaceae

Description: Broad-leaved deciduous shrub or tree, up to 40 feet tall; smooth dark bluish gray bark with musclelike ridges; simple, double-toothed, egg-shaped leaves (up to 5 inches long), alternately arranged; inconspicuous flowers borne on catkins; ribbed fruit nutlets.

Flowering period: March and April.

Habitat: Tidal swamps; nontidal temporarily flooded forested wetlands along floodplains, hydric hammocks, moist woods, and bottomlands.

Wetland indicator status: FAC.

Range: Nova Scotia to Minnesota, south to Florida and Texas.

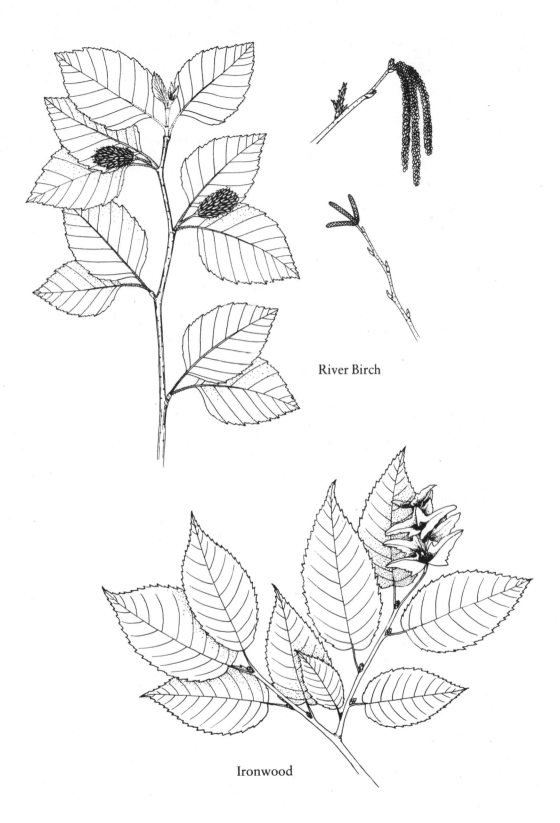

River Birch

Ironwood

Black Willow

Salix nigra Marshall

Willow Family
Salicaceae

Description: Deciduous shrub or tree, up to 70 feet tall or more; trunk up to 20 inches in diameter with brownish to blackish, deeply grooved bark; yellow-brown to dark brown branchlets, often hairy when young; simple, narrowly lance-shaped, fine-toothed leaves (up to 5 inches long and to ⅖ inch wide) tapering to a long point, green above, light green below, leaf stalks (to ⅖ inch long) often hairy, alternately arranged, with somewhat heart-shaped toothed leaflike structures (stipules, up to ½ inch long) at leaf bases; inconspicuous flowers borne on dense spikes (catkins) at end of short, leafy peduncles; somewhat pear-shaped fruit capsule.

Flowering period: March and April.

Habitat: Irregularly flooded tidal fresh marshes and swamps; nontidal swamps, marshes, floodplain forests, wet meadows, streambanks, and river bars.

Wetland indicator status: OBL.

Range: Southern Canada to central Minnesota, south to Florida and Texas.

Similar species: Swamp or Carolina Willow (*S. caroliniana*) is quite similar but has leaves with whitish undersides (usually slightly hairy) and longer leaf stalks (to ⅖ inch long); it is a shrub or small tree; it is OBL.

American Elm

Ulmus americana L.

Elm Family
Ulmaceae

Description: Broad-leaved deciduous tree, up to 125 feet tall; gray, scaly ridged bark; simple, coarse-toothed leaves (up to 6 inches long) with unequal leaf bases, smooth or slightly rough above and smooth or hairy below, alternately arranged; inconspicuous greenish flowers borne in clusters on long drooping stalks; elliptical flattened winged fruits (to ½ inch long).

Flowering period: February through March.

Habitat: Tidal swamps; nontidal forested wetlands, especially along floodplains, hydric hammocks, and moist rich upland woods.

Wetland indicator status: FACW.

Range: Newfoundland to Manitoba, south to Florida and Texas.

Similar species: Water Elm or Planer Tree (*Planera aquatica*), a shrub or short tree (usually less than 30 feet high), has leaves with slightly unequal bases, but they are smaller (to about 2½ inches long), narrower (to 1 inch wide), and tapered above and widest below the middle; it is OBL.

SEE ALSO Water Gum (*Nyssa aquatica*), described under Black Gum (*N. sylvatica*).

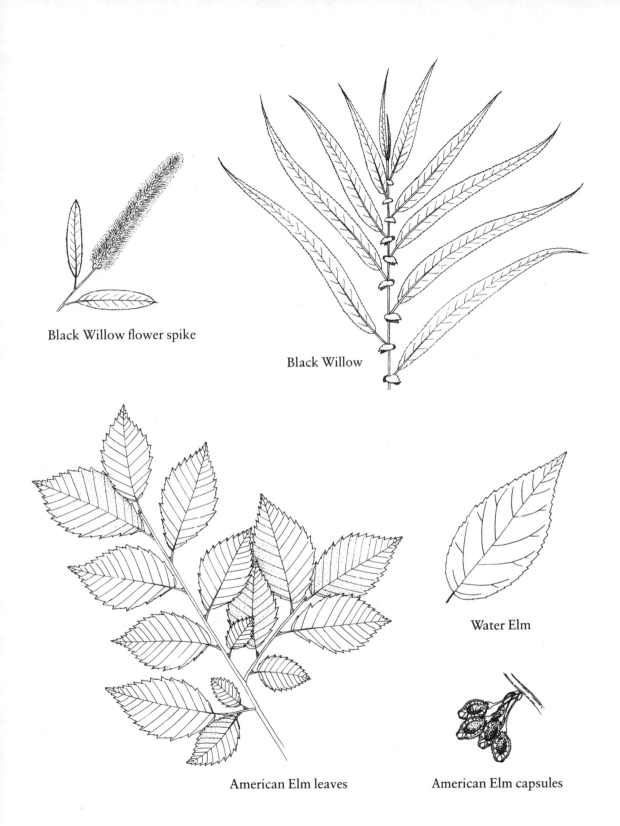

Black Willow flower spike

Black Willow

Water Elm

American Elm leaves

American Elm capsules

DECIDUOUS TREES WITH SIMPLE, TOOTHED, LOBED ALTERNATE LEAVES

Overcup Oak

Quercus lyrata Walter

Beech Family
Fagaceae

Description: Broad-leaved deciduous tree, up to 100 feet tall; light gray scaly ridged bark; twigs brown to gray; simple deeply lobed leaves (up to 10 inches long) with three to fourteen lobes, shiny dark green above and often hairy below, alternately arranged; inconspicuous flowers borne on catkins (male) and singly or in pairs (female); acorn (½–1 inch long) nearly completely covered by cup.

Flowering period: March and April.

Habitat: Tidal swamps; nontidal forested wetlands, especially along floodplains.

Wetland indicator status: OBL.

Range: Southern New Jersey to Florida and Texas, north to Illinois and Indiana.

Sweet Gum

Liquidambar styraciflua L.

Witch Hazel Family
Hamamelidaceae

Description: Broad-leaved deciduous tree, up to 140 feet tall; deeply ridged gray bark; twigs often with corky wings; simple, five- to seven-lobed (star-shaped) toothed leaves (up to 8 inches long), shiny green above, aromatic when crushed, alternately arranged; inconspicuous greenish flowers borne in ball-like clusters; hard spiny fruit balls.

Flowering period: April through October.

Habitat: Tidal swamps; nontidal forested wetlands, hydric hammocks, moist upland woods, clearings, and old fields.

Wetland indicator status: FAC+.

Range: Southern Connecticut to southern Illinois and Oklahoma, south to Florida and Mexico.

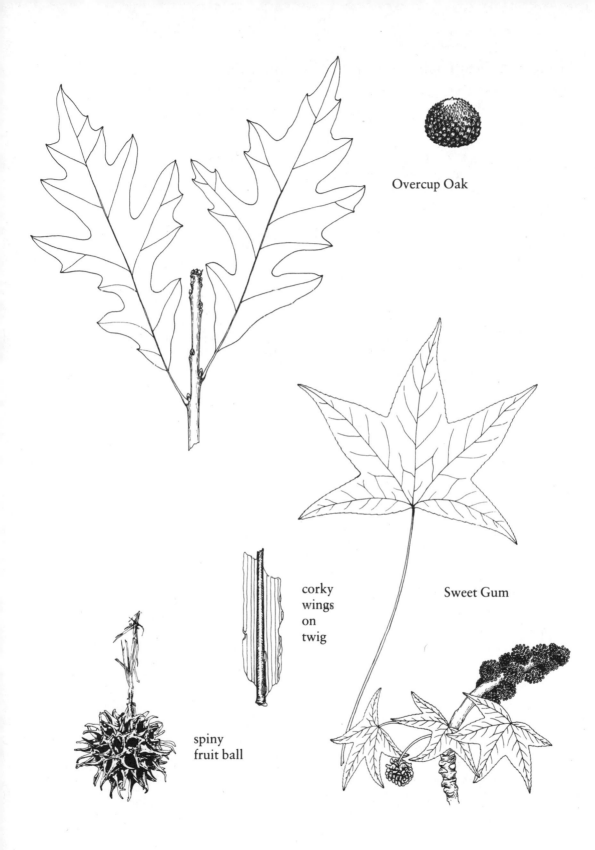

Overcup Oak

corky
wings
on
twig

Sweet Gum

spiny
fruit ball

Tulip or Yellow Poplar

Liriodendron tulipifera L.

Magnolia Family
Magnoliaceae

Description: Broad-leaved deciduous tree, up to 190 feet tall; smooth light gray young bark and deeply grooved darker older bark; simple, toothed, distinctly four-lobed (sometimes six-lobed) leaves (up to 10 inches long), shiny dark green above, long-stalked, aromatic when crushed, alternately arranged; large, showy, tuliplike, six-petaled greenish yellow flowers (1½–2 inches wide) borne singly on end of twigs; conelike fruit capsules (to 3 inches long). (*Note:* Fruit capsules persist through winter.)

Flowering period: April through June.

Habitat: Tidal swamps (at higher elevations); nontidal forested wetlands, hydric hammocks, rich moist upland woods, and abandoned fields.

Wetland indicator status: FAC.

Range: Vermont and southern Ontario to Michigan and Missouri, south to Florida and Louisiana.

Sycamore

Platanus occidentalis L.

Sycamore Family
Platanaceae

Description: Broad-leaved deciduous tree up to 175 feet tall; flaky bark mottled with gray, green, brown, white, and yellow, peeling off in large flakes; simple, three- to five-lobed, large-toothed leaves (up to 10 inches long) alternately arranged; inconspicuous greenish flowers borne in ball-like clusters; round fruit balls (to 1½ inches wide) hanging on long stalks.

Flowering period: March through June.

Habitat: Tidal swamps (at higher elevations); nontidal temporarily flooded forested wetlands along floodplains, moist alluvial woods, bottomlands, edges of lakes and swamps, and dry slopes.

Wetland indicator status: FACW−.

Range: Southern Maine to Minnesota, south to Florida and Texas.

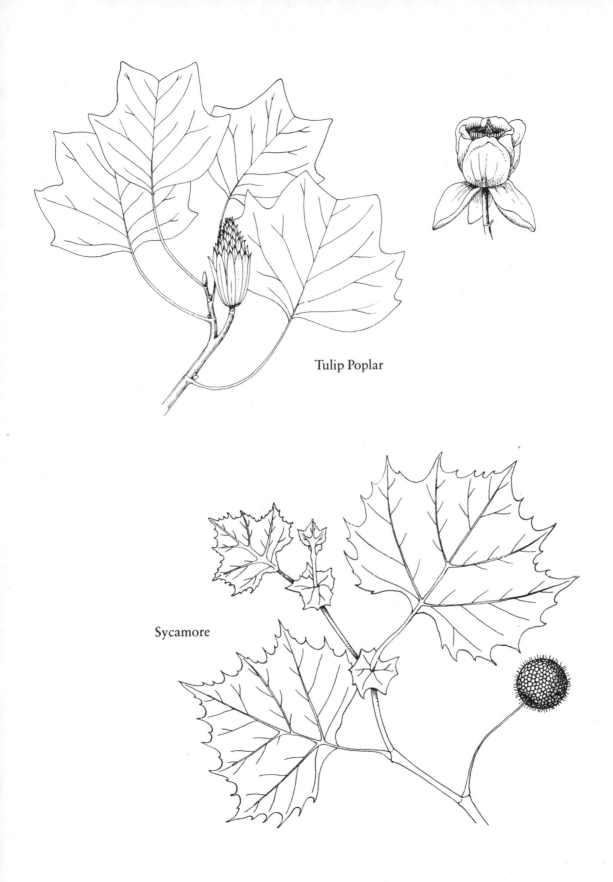

Tulip Poplar

Sycamore

HERBACEOUS VINES WITH REDUCED LEAVES

Common Dodder

Cuscuta gronovii Willd. ex J. A. Schultes

Morning Glory Family
Convolvulaceae

Description: Slender, parasitic, twining, herbaceous nonleafy vines; stems smooth and orange or yellow-orange; leaves apparently lacking but actually reduced to minute scales; small white or yellowish bell-shaped flowers ($\frac{1}{10}$–$\frac{1}{5}$ inch long) with rounded lobes borne in sessile cluster; ball-shaped fruit capsule.

Flowering period: July to October.

Habitat: Tidal fresh marshes, occasionally salt and brackish marshes, especially parasitizing High-tide Bush (*Iva frutescens*); nontidal marshes and other low areas.

Wetland indicator status: Not designated, since plant is an epiphyte.

Range: Nova Scotia to Manitoba, south to Florida, Texas, and Arizona.

Similar species: Distinguishing Common Dodder from other dodders requires close examination of flowers and use of a technical taxonomic reference. Pretty Dodder (*C. indecora*) occurs in southern salt and brackish marshes and freshwater wetlands; its flowers have pointed lobes.

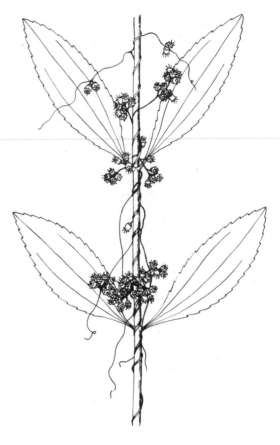

Common Dodder

HERBACEOUS VINES WITH COMPOUND LEAVES

Ground-nut

Apios americana Medic.

Legume Family
Leguminosae

Description: Twining perennial vine; rhizomes with two or more tubers; compound leaves divided into five to seven (sometimes three) lance-shaped leaflets (1⅗–2⅖ inches long) tapering to a fine point distally and somewhat rounded at base; irregular five-petaled (somewhat two-lipped) purplish or brownish fragrant pealike flowers (about ½ inch long) borne singly or in pairs on dense axillary clusters (racemes); linear fruit pod (up to 4 inches long).

Flowering period: June into September.

Habitat: Irregularly flooded tidal fresh marshes; edges of nontidal swamps, moist thickets and woods, and borders of streams or ponds.

Wetland indicator status: FACW.

Range: Quebec to Minnesota and South Dakota, south to Florida and Texas.

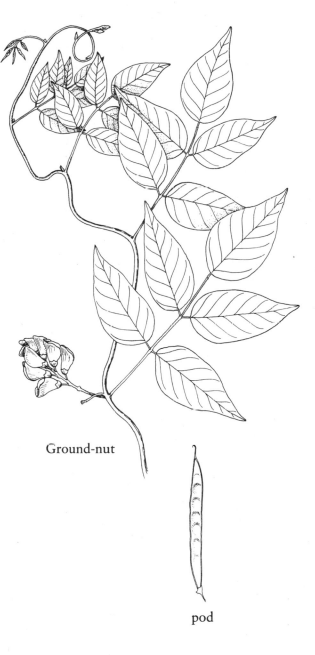

Ground-nut

pod

Leather-flower
or Swamp Virgin's Bower

Clematis crispa L.

Buttercup Family
Ranunculaceae

Description: Herbaceous vine with smooth, many-angled stem; compound leaves with mostly three to five entire or lobed, linear to oval-shaped leaflets (up to 4 inches long and to 2 inches wide), somewhat short-stalked, oppositely arranged; showy, four-"petaled" bluish, rose, or whitish flowers (more than 1 inch wide) borne singly at end of branch, "petals" somewhat spongy with wrinkled margins and a long narrow tip; one-seeded dry fruit (about ⅓ inch wide).

Flowering period: April to August.

Habitat: Tidal fresh marshes; nontidal marshes and swamps, hydric hammocks, wet pine flatwoods and savannahs, river and stream banks, and wet thickets.

Wetland indicator status: FACW+.

Range: Southeastern Virginia to central Florida, west to eastern Texas, and north to southern Illinois.

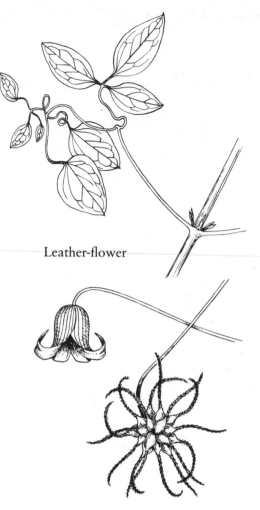

Leather-flower

fruit cluster

Climbing Hempweed

Mikania scandens (L.) Willd.

Composite or Aster Family
Compositae (*Asteraceae*)

Description: Twining and sprawling herbaceous vine, up to 20 feet long; stems hairy to nearly smooth; simple, slightly toothed leaves (1–5½ inches long), somewhat triangle-shaped with rounded bases or heart-shaped, tapering to a slender point, with three main veins, oppositely arranged; numerous white or pink flowers in heads borne in stalked clusters arising from leaf axils.

Flowering period: July through October.

Habitat: Tidal fresh marshes; nontidal marshes, swamps, hydric hammocks, and stream banks.

Wetland indicator status: FACW+.

Range: Southern Maine to Florida and Texas; locally inland to Michigan and Missouri.

Similar species: Ladies'-eardrops (*Brunnichia cirrhosa*) has similarly shaped leaves that are alternately arranged; it climbs by tendrils from lateral branches, has greenish white to greenish five-lobed tubular flowers borne in paniclelike spikes, and has distinctive elongate, winged showy pink fruits; it is FACW.

Climbing Hempweed

Hedge Bindweed

Calystegia sepium (L.) R. Br.
[*Convolvulus sepium* L.]

Morning Glory Family
Convolvulaceae

Description: Twining and sometimes trailing herbaceous vine, up to 10 feet long; simple, entire, triangular-shaped leaves (2–4 inches long), often with somewhat squarish basal lobes, on long petioles, alternately arranged; large white, pink, or purplish funnel-shaped tubular flowers (1½–3 inches long) borne usually singly on long stalks (peduncles, 2–6 inches long).

Flowering period: May into October.

Habitat: Tidal fresh marshes, occasionally brackish marshes and beaches; inland moist thickets, edges of nontidal marshes, shores, roadsides, and waste places.

Wetland indicator status: FAC.

Range: Quebec and Newfoundland to British Columbia, south to Florida, Missouri, and Oregon.

Similar species: See Salt Marsh Morning Glory (*Ipomoea sagittata*).

Wild Yam

Dioscorea villosa L.

Yam Family
Dioscoreaceae

Description: Twining herbaceous perennial vine, up to 16 feet long; rhizomes tuberous; stems smooth; simple, entire, heart-shaped leaves (2–4 inches long) tapering distally to a slender curved point, with seven to eleven prominent veins, petioled, alternately arranged; small white or greenish yellow flowers borne on short spikes (up to 4 inches long), female flowers borne singly along spike and male flowers borne singly or in clusters of up to four; three-winged fruit capsule.

Flowering period: April through November.

Habitat: Irregularly flooded tidal fresh marshes and swamps; nontidal shrub swamps, forested wetlands, and roadsides.

Wetland indicator status: FAC.

Range: Connecticut to Minnesota, south to Florida and Texas.

Hedge Bindweed

Wild Yam

WOODY VINES ARMED WITH THORNS OR PRICKLES

Common Greenbrier

Smilax rotundifolia L.

Lily Family
Liliaceae

Description: Climbing woody vine forming dense tangles; round or square many-branched stems with many stout thorny prickles (green bases and dark tips); simple, entire, broadly heart-shaped or rounded leaves (up to 5 inches long), shiny green above, alternately arranged; inconspicuous greenish or bronze-colored flowers borne in clusters; bluish black to black berries.

Flowering period: April into June.

Habitat: Tidal swamps; nontidal forested wetlands (especially temporarily flooded), margins of wooded swamps, open upland woods and thickets, and roadsides.

Wetland indicator status: FAC.

Range: Nova Scotia to Michigan, south to Florida and Texas.

Similar species: Bullbrier or Saw Greenbrier (*S. bona-nox*) also has black berries but has leathery triangular leaves often fringed with bristles; it is FAC. See Bamboo Vine or Laurel-leaved Greenbrier (*S. laurifolia*).

Red-berried Greenbrier

Smilax walteri Pursh

Lily Family
Liliaceae

Description: Low, scrambling woody vine; round to weakly angled stems with slender prickles (sometimes stout thorny) on lower part, mostly smooth branches, and many tendrils; simple, entire leaves (up to 5 inches long), shiny green above, alternately arranged; inconspicuous greenish or bronze-colored flowers borne in clusters; bright red berries.

Flowering period: April through May.

Habitat: Tidal swamps; nontidal forested wetlands and bogs.

Wetland indicator status: OBL.

Range: New Jersey to Florida, Texas, and western Tennessee.

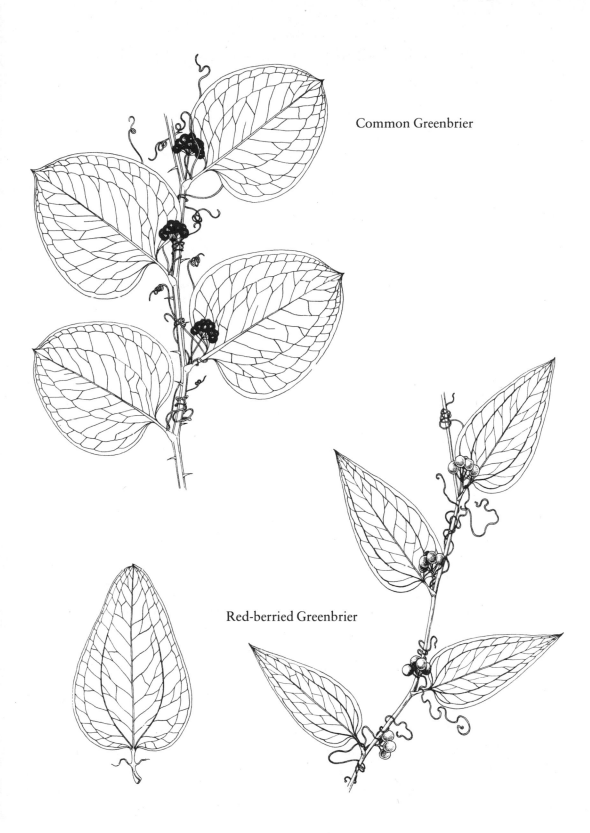

Common Greenbrier

Red-berried Greenbrier

Laurel-leaved Greenbrier or **Bamboo Vine**

Smilax laurifolia L.

Lily Family
Liliaceae

Description: High-climbing woody vine; stem without spiny prickles except near bases and with few or no tendrils; simple, entire, leathery, evergreen narrowly oblong leaves (up to 8 inches long) with pointed tips, alternately arranged; inconspicuous greenish yellow flowers borne in clusters; black berries.

Flowering period: July into September.

Habitat: Tidal swamps; nontidal seasonally flooded forested wetlands, pocosins, cypressgum swamps, wet pine flatwoods, and pond borders.

Wetland indicator status: FACW+.

Range: New Jersey to Florida, Texas, and Arkansas; also in the Bahamas and Cuba.

Similar species: Ear-leaf Greenbrier (*S. auriculata*) has been reported in Florida salt marshes, although it is primarily a dune species; its stem usually lacks spiny prickles, but its evergreen leaves are oblong with rounded or wedge-shaped bases (sometimes distinctly lobed basally); it is FACU.

flower

Laurel-leaved Greenbrier

Poison Ivy

Toxicodendron radicans (L.) Kuntze
[*Rhus radicans* L.]

Cashew Family
Anacardiaceae

Description: Erect broad-leaved deciduous
shrub, trailing vine, or climbing plant, up to 10
feet tall when not climbing; twigs brown, older
climbing stems densely covered by dark fibers;
sap milky; long-stalked compound leaves (4–14
inches long) divided into three leaflets, end leaflet having a longer stalk than side leaflets, alternately arranged; small yellowish flowers with
five petals borne on lateral clusters (panicles, up
to 4 inches long); small grayish white fruit balls
borne in clusters. (*Warning: Do not touch*; plant
is poisonous and may cause severe skin irritations.)

Flowering period: April into June.

Fruiting period: August through November
(mostly); some persist through winter.

Habitat: Tidal fresh marshes and swamps, and
along the upper edges of salt marshes; various
habitats, mostly dry woods and thickets, but
also common in nontidal wetlands, including
hydric hammocks.

Wetland indicator status: FAC.

Range: Nova Scotia and Quebec to British Columbia, south to Florida and Mexico.

Similar species: Poison Sumac (*T. vernix*, formerly *Rhus vernix*) occurs in tidal fresh marshes
and swamps; it is a shrub or small tree (6–23
feet tall) with leaves divided into seven to thirteen leaflets; it is OBL.

Poison Ivy

Cross Vine

Bignonia capreolata L.

Trumpet Creeper Family
Bignoniaceae

Description: Climbing woody vine, up to 50 feet
long; oppositely arranged compound leaves divided into two entire, stalked, oblong to egg-
shaped leaflets (up to 6 inches long) with somewhat heart-shaped bases, tendrils arising from
between leaflets; reddish or orange five-lobed,
bell-shaped tubular flowers (up to 2 inches long)
with yellow or red inner parts, in two to five
clusters borne in leaf axils; flattened capsules
(to 8 inches long). (*Note:* Leaves turn purplish
color in winter.)

Flowering period: March through May.

Habitat: Tidal swamps; nontidal seasonally
flooded forested wetlands, hydric hammocks,
alluvial forests, rich woods, and thickets.

Wetland indicator status: FAC.

Range: Southern Maryland to southern Missouri, south to Florida and Louisiana.

Trumpet Creeper

Campsis radicans (L.) Seem.

Trumpet Creeper Family
Bignoniaceae

Description: Trailing or climbing woody vine,
up to 30 feet or more long; aerial rootlets in
two rows along twigs; oppositely arranged compound leaves (up to 12 inches long) divided into
five to thirteen coarse-toothed leaflets (up to 3¼
inches long); several large tubular five-lobed,
orange-red flowers (to 3¼ inches long) borne in
terminal clusters; podlike fruit capsules (to 6
inches long).

Flowering period: May through October.

Habitat: Tidal swamps; nontidal forested wetlands, hydric hammocks, moist upland woods,
thickets, fence rows, and roadsides.

Wetland indicator status: FAC.

Range: Connecticut to Michigan and Iowa,
south to Florida and Texas; also farther north
when escaped from cultivation.

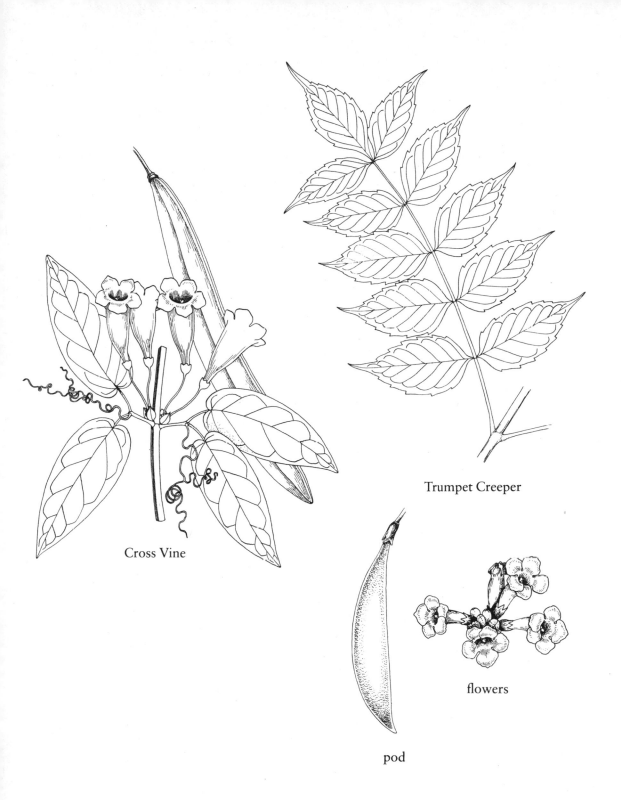

Cross Vine

Trumpet Creeper

pod

flowers

Pepper-vine

Ampelopsis arborea (L.) Koehne

Grape Family
Vitaceae

Description: Climbing woody vine, sometimes an erect shrub; stems smooth with white pith; alternately arranged compound leaves (up to 6 inches long or more) twice or thrice divided into numerous toothed, variably sized leaflets (up to 2½ inches long and to 1¾ inches wide) with mostly wedge-shaped bases, dark green above, light green below; many small greenish to yellowish five-petaled flowers borne in clusters (cymes, less than 3¼ inches long) on long stalks from stem opposite the leaves; somewhat roundish black berries (about ½ inch wide).

Flowering period: June through October.

Habitat: Tidal swamps and fresh marshes and upper edges of brackish marshes; nontidal marshes and swamps, hydric hammocks, woodland borders, fence rows, and waste places.

Wetland indicator status: FAC+.

Range: Maryland and Illinois south to Florida, Texas, and eastern New Mexico.

Similar species: Marine Ivy (*Cissus incisa*), a related vine species with warty stems and light green, fleshy, three-parted (simple or compound) toothed leaves (1–3 inches long), occurs in salt marshes and maritime forests along the Gulf coast; it is FAC.

Virginia Creeper

Parthenocissus quinquefolia (L.) Planch.

Grape Family
Vitaceae

Description: Trailing or climbing, soft woody vine, climbing by adhesive disks at tips of tendrils; alternately arranged compound leaves divided into three to seven (usually five) coarse-toothed leaflets; small greenish or whitish flowers borne in terminal umbellike clusters; dark blue to blackish berries.

Flowering period: May into August.

Habitat: Tidal swamps; nontidal shrub wetlands, forested wetlands, hydric hammocks, rich upland woods, moist thickets, and fence rows.

Wetland indicator status: FAC.

Range: Maine to Ohio and Kansas, south to Florida and Texas.

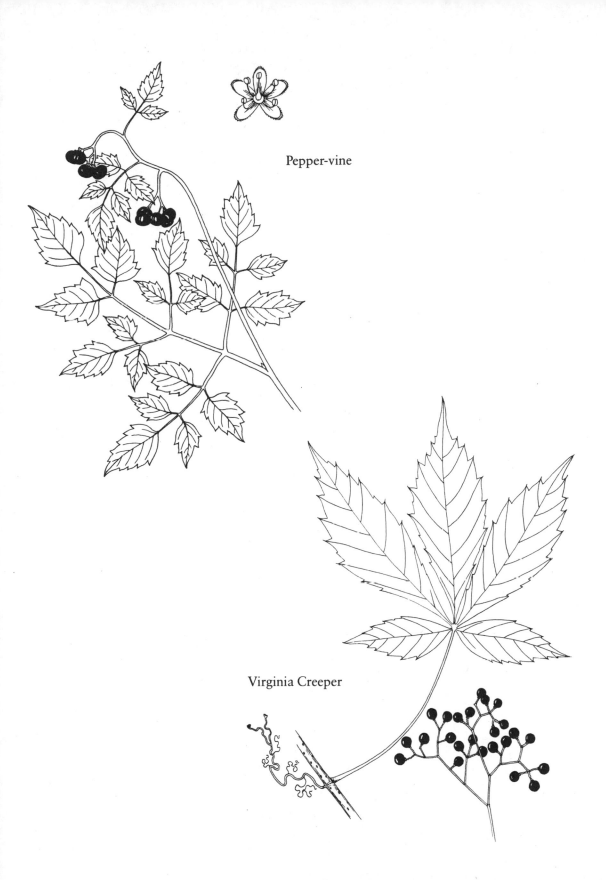

Pepper-vine

Virginia Creeper

Distribution of Coastal Wetlands in the Southeast

More than 80 percent of the nation's coastal wetlands (in the lower forty-eight states) are found in the southeastern states. The following discussion briefly describes the general distribution of these wetlands in each state and lists some places to observe coastal wetlands. Table 3 summarizes the acreages of coastal wetlands for these states. The general locations of coastal marshes are shown on an accompanying map for each state. More detailed wetland maps are available from the U.S. Fish and Wildlife Service's National Wetlands Inventory Project (to order call 1-800-USA-MAPS). Other coastal wetland maps may be available from state agencies.

Coastal wetlands are easily seen from highways, causeways, and bridges leading to the beaches and coastal communities. For a closer look, I recommend a visit to a nearby refuge, wildlife management area, park, nature sanctuary, or natural preserve owned by federal, state, or local agencies or by private nonprofit organizations. These conservation areas encourage public use and often have interpretive trails through or beside the wetlands. Some may even have guided tours by park naturalists, so contact them before you visit.

Some coastal wetlands open to the public are listed for each state. Undoubtedly, there are others to visit; for information on their location, contact the local parks department or planning office.

Virginia

Virginia may possess about 210,000 acres of coastal marshes and about 70,000 acres of tidal swamps. Coastal marshes are most abundant along Virginia's Eastern Shore in Accomac and Northampton Counties, where they have formed behind numerous barrier islands facing the Atlantic Ocean or along Chesapeake Bay. Elsewhere in the state, coastal wetlands have mostly developed along tidal rivers, such as the James, York, Rappahannock, and Potomac. For more information on Virginia's tidal wetlands, contact the Virginia Institute of Marine Science at Gloucester Point, which has published a series of atlases of tidal wetland maps for each county.

Several places to see coastal wetlands include (1) Chincoteague National Wildlife Refuge (Chincoteague), (2) Assateague National Seashore (Assateague Island), (3) Eastern Shore National Wildlife Refuge (Cape Charles), (4) Back Bay National Wildlife Refuge (Virginia Beach), (5) Mason Neck National Wildlife Refuge (Woodbridge), (6) Presquille National Wildlife

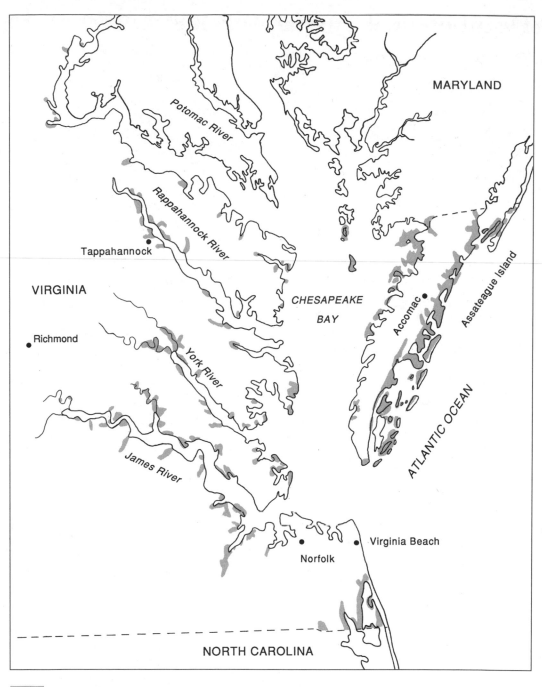

MARYLAND

Potomac River

Rappahannock River

Tappahannock

VIRGINIA

CHESAPEAKE
BAY

Accomac

Assateague Island

Richmond

York River

ATLANTIC OCEAN

James River

Virginia Beach

Norfolk

NORTH CAROLINA

Coastal Marshes

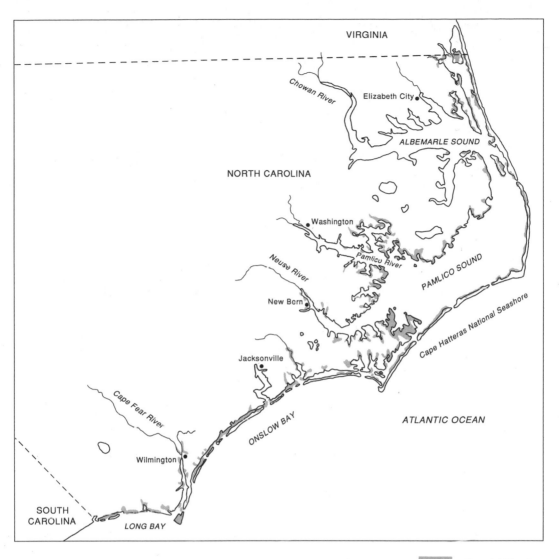

VIRGINIA

Chowan River

Elizabeth City

ALBEMARLE SOUND

NORTH CAROLINA

Washington

Pamlico River

PAMLICO SOUND

Neuse River

New Bern

Cape Hatteras National Seashore

Jacksonville

Cape Fear River

ONSLOW BAY

ATLANTIC OCEAN

Wilmington

SOUTH
CAROLINA

LONG BAY

Coastal Marshes

Distribution of Coastal Wetlands in the Southeast   301

Table 3. Estimated acreage of coastal wetland types in southeastern states

State	Estuarine Marshes (*acres*)	Estuarine Swamps (*acres*)	Tidal Fresh Marshes (*acres*)	Tidal Swamps (*acres*)	Tidal Flats (*acres*)
Virginia	184,100	6,600	20,700	68,700	107,200
North Carolina	212,800	29,700	2,200	3,500	44,000
South Carolina	365,900	3,400	46,300	101,800	32,000
Georgia	194,100	2,400	8,600	58,700	7,500
Florida/South Atlantic	106,000	66,800	6,500	10,500	16,100
Florida/Gulf Coast	257,200	613,800	9,800	18,400	193,900
Florida (total) ·	363,200	680,600	16,300	28,900	210,000
Alabama	25,500	2,700	200	2,000	4,200
Mississippi	58,900	1,000	—	—	2,300
Louisiana	1,722,900	10,200	65,000	4,800	31,800
Texas	432,100	2,900	22,700	7,500	275,300

Source: Field et al. *Coastal Wetlands of the United States.*
Note: Estuarine swamps include mangrove swamps and other estuarine wetlands dominated by halophytic shrubs.

Refuge (Hopewell), (7) Leesylvania State Park (Woodbridge), (8) York River State Park (Croaker/Williamsburg), (9) False Cape State Park (Sandbridge/Virginia Beach), (10) Westmoreland State Park–Big Meadows Swamp (Montross), (11) Chippokes Plantation State Park (Surry), and (12) Virginia Marine Science Museum (Virginia Beach).

North Carolina

North Carolina's coastal marshes may total about 245,000 acres, and about 4,000 acres of tidal swamps may exist. The state's coastline is dominated by a series of offshore islands forming the Outer Banks. These islands protect the mainland and restrict the flow of salt water into North Carolina's estuaries. Consequently, most of the coastal marshes are brackish marshes dominated by black needlerush. These marshes occur along Pamlico and Albemarle Sounds and upstream along tidal rivers. Salt marshes have developed along the bay side of the barrier islands and near the inlets of vari-

ous sounds (e.g., Bogue and Back Sounds). Salt and brackish marshes are most abundant in Carteret and Hyde Counties.

Coastal wetlands can be studied closely at the following locations: (1) Mackay Island National Wildlife Refuge (Knotts Island), (2) Pea Island National Wildlife Refuge (Manteo), (3) Cedar Island National Wildlife Refuge (Swanquarter), (4) Cape Lookout National Seashore–Core Banks and Portsmouth Island, (5) Cape Hatteras National Seashore–Ocracoke Island, (6) Rachel Carson National Estuarine Research Reserve (Beaufort), (7) North Carolina Aquarium–Roanoke Island (Manteo), (8) North Carolina Aquarium–Pine Knoll Shores, (9) Marine Resource Center–Fort Fisher (Kure Beach), (10) Goose Creek State Park (Washington), (11) Cedar Point Recreation Area (Swansboro), and (12) Lawson Creek Park (New Bern).

South Carolina

South Carolina may have over 400,000 acres of coastal marshes and about

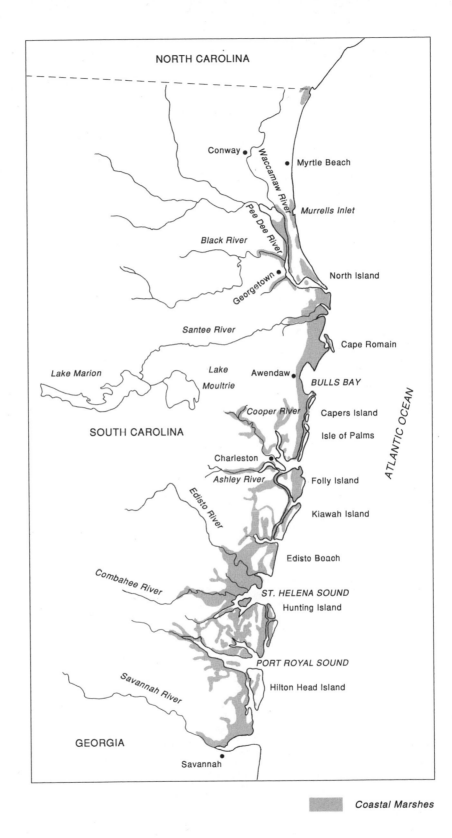

NORTH CAROLINA

Conway •

Waccamaw River

• Myrtle Beach

Murrells Inlet

Pee Dee River

Black River

Georgetown •

North Island

Santee River

Cape Romain

Lake Marion

Lake Moultrie

Awendaw •

BULLS BAY

SOUTH CAROLINA

Cooper River

Capers Island

Isle of Palms

Charleston •

Ashley River

Folly Island

Edisto River

Kiawah Island

Edisto Beach

Combahee River

ST. HELENA SOUND

Hunting Island

PORT ROYAL SOUND

Savannah River

Hilton Head Island

GEORGIA

Savannah •

ATLANTIC OCEAN

Coastal Marshes

Distribution of Coastal Wetlands in the Southeast 303

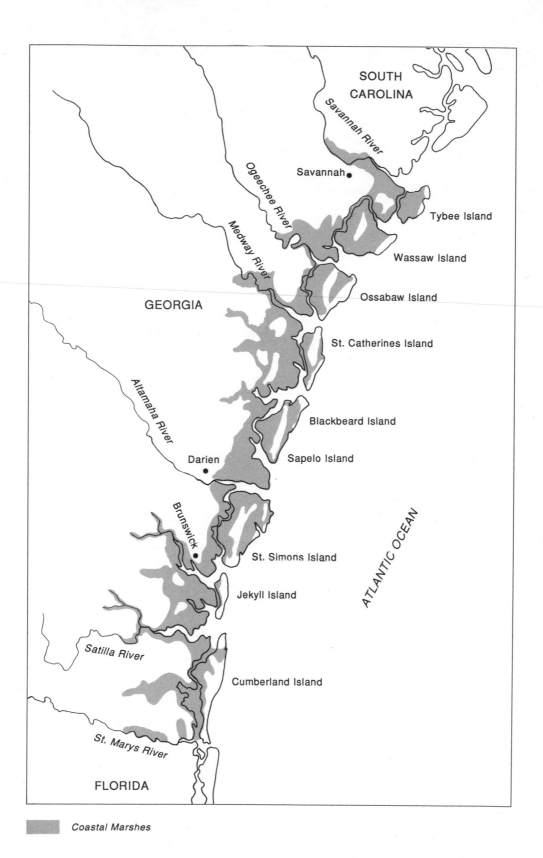

SOUTH
CAROLINA

Savannah River

Savannah•

Tybee Island

Ogeechee River

Wassaw Island

Medway River

GEORGIA

Ossabaw Island

St. Catherines Island

Altamaha River

Blackbeard Island

Darien•

Sapelo Island

Brunswick•

ATLANTIC OCEAN

St. Simons Island

Jekyll Island

Satilla River

Cumberland Island

St. Marys River

FLORIDA

Coastal Marshes

100,000 acres of tidal swamps. Salt marshes dominated by smooth cordgrass occur behind the barrier islands forming the state's coastline from North Inlet at the mouth of Winyah Bay to the Georgia border. They are most abundant in Charleston and Beaufort Counties. Upstream along tidal rivers, such as the Santee, Cooper, Ashley, and Stono, brackish marshes are found. Tidal fresh marshes and swamps are common along coastal rivers with large drainage basins, including the Pee Dee, Waccamaw, Santee, Cooper, Edisto, Combahee, Ashepoo, and Savannah. More detailed descriptions of South Carolina's coastal marshes can be found in *An Inventory of South Carolina's Coastal Marshes*, available from the South Carolina Wildlife and Marine Resources Department in Charleston.

Coastal marshes can be seen at the following refuges and parks: (1) Cape Romain National Wildlife Refuge (Awendaw), (2) Pinckney Island National Wildlife Refuge (near Savannah, GA), (3) Huntington Beach State Park (Murrells Inlet), (4) Edisto Beach State Park (Edisto Island), (5) Hunting Island State Park (St. Helena Island), and (6) Folly Beach County Park (Folly Beach/Charleston).

Georgia

Coastal marshes may occupy about 200,000 acres in Georgia, and nearly 60,000 acres of tidal swamps may be present. A band of salt marshes several miles wide lies between the mainland and the offshore barrier islands called "sea islands." Among these islands is perhaps the most famous, Sapelo Island, where the University of Georgia's Marine Institute is located. This center and its neighboring salt marshes are the foundation for much of our knowledge about U.S. coastal wetlands. Brackish and tidal fresh wetlands occur along numerous rivers draining the Coastal Plain, including the Savannah, Ogeechee, Altamaha, Satilla, and St. Marys. Salt and brackish marshes are most common in Chatham, Camden, McIntosh, and Glynn Counties.

The following places provide public access to coastal marshes: (1) Savannah National Wildlife Refuge (Savannah), (2) Cumberland Island National Seashore (St. Marys), (3) Skidaway Island State Park (Savannah), (4) Crooked River State Park (St. Marys), (5) Sapelo Island Sanctuary (Darien), and (6) Oatland Island Education Center (Savannah).

Florida

Largely a peninsula separating the Atlantic Ocean from the Gulf of Mexico, Florida may possess roughly 1.3 million acres of coastal wetlands, with about 380,000 acres of salt and brackish marshes, 680,000 acres of mangrove swamps, and about 30,000 acres of tidal freshwater swamps. The marshes are most abundant along the Gulf coast north of Tarpon Springs to Apalachee Bay (St. Marks National Wildlife Refuge). Here brackish marshes of black needlerush extend directly into Gulf waters without the protection of barrier islands due to little or no tidal energy. Coastal marshes are most abundant in Monroe County. Unique among other southeastern states, Florida has many acres of mangrove swamps. These saltwater swamps are most extensive from Naples to Miami, comprising the "Ten Thousand Islands" and the southern edge of Everglades National Park.

There are many places to see coastal wetlands, including (1) Merritt Island National Wildlife Refuge (Titusville), (2) Ding Darling National Wildlife Refuge (Sanibel), (3) St. Marks National Wildlife Refuge (St. Marks), (4) Chassahowitzka National

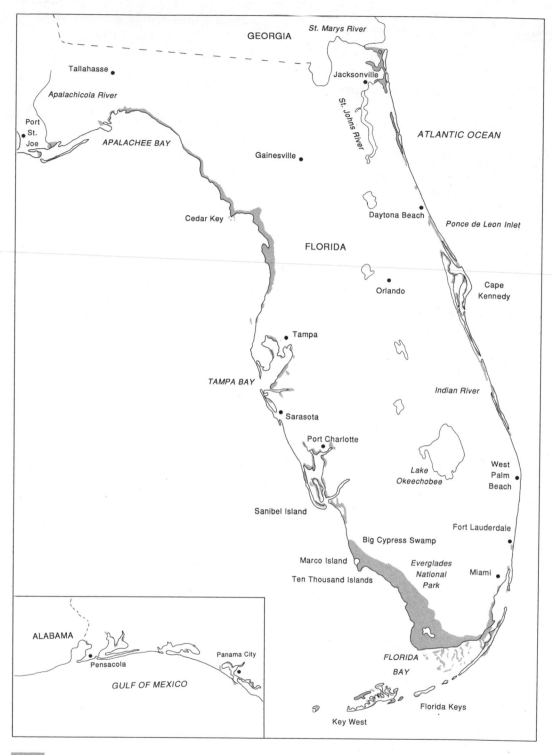

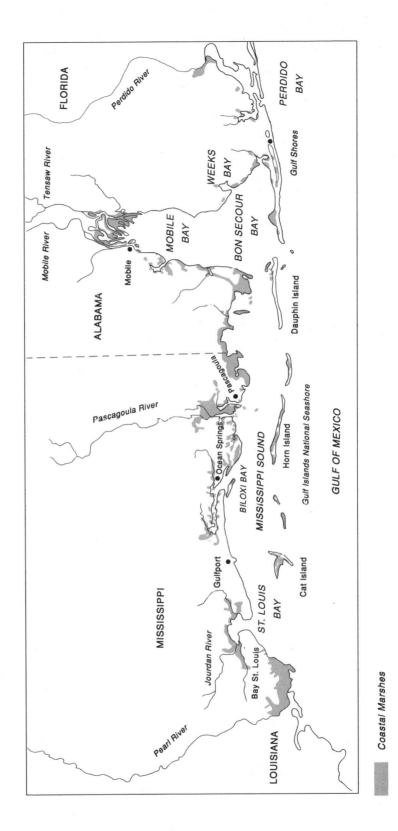

Wildlife Refuge (Homosassa), (5) Everglades National Park (Flamingo and Everglades City), (6) Rookery Bay National Estuarine Sanctuary (Shell Island), (7) Gulf Islands National Seashore—Perdido Key (Gulf Breeze), (8) Anastasia State Park (St. Augustine Beach), (9) Tomoka State Park (Ormond Beach), (10) Jonathan Dickinson State Park (Jupiter), (11) Coastal Science Center (Stuart/Hutchinson Island), (12) Collier-Seminole State Park (Naples), (13) Honeymoon Island State Park (Dunedin), (14) State Park and Waccasassa Bay State Preserve (Cedar Key), (15) Big Lagoon State Recreation Area (Pensacola), (16) Sanibel-Captiva Conservation Foundation Nature Trails (Sanibel), (17) Hillsborough Community College Environmental Studies Center (Sun City), (18) Upper Tampa Bay Park (Tampa), (19) Hammock Park (Dunedin), and (20) University of West Florida Campus Trail (Pensacola).

Alabama

Alabama's coastline occupies only the southwestern corner of the state. Dominant coastal-zone features include Mobile Bay, Bon Secour Bay, Mississippi Sound, Perdido Bay, Dauphin Island, and Little Lagoon. Coastal marshes may represent about 25,000 acres and are most extensive in Mobile County. Salt and brackish marshes dominated by black needlerush are common behind barrier islands and beaches and along the shores of sheltered embayments. Tidal fresh wetlands are best represented along the Mobile, Tensaw, and Perdido Rivers.

A few places to experience coastal wetlands are (1) Weeks Bay National Estuarine Research Reserve (Fairhope), (2) Bon Secour National Wildlife Refuge (Gulf Shores), and (3) Meaher State Park (Mobile).

Mississippi

Mississippi's mainland is protected by a string of barrier islands that separate Mississippi Sound from the Gulf of Mexico. These barriers provide a favorable environment for the establishment of coastal wetlands behind the islands and along the shores of Mississippi Sound. Coastal wetlands also form along embayments, such as St. Louis Bay, Biloxi Bay, and Pascagoula Bay, and along tidal rivers, including the Jourdan, Pearl, and Pascagoula. Roughly 60,000 acres of coastal marshes may exist, with black needlerush dominating over 90 percent of them.

Coastal marshes can be visited at (1) Gulf Island National Seashore—Davis Bayou (Ocean Springs), (2) Buccaneer State Park (Waveland), and (3) Shepard State Park (Gautier).

Louisiana

Louisiana leads the nation in coastal wetland acreage with about 2 million acres. These wetlands create a marshy belt, typically more than 10 miles wide, that forms the state's coastline. Past and present deltas of the Mississippi River occupy about two-thirds of the state's coastal zone, and the rest comprises the Chenier Plain that formed by deposition of river sediments carried west by Gulf currents. Consequently, the Mississippi River and its sediments are responsible for Louisiana's vast wetland acreage. Despite their abundance, Louisiana's coastal marshes are disappearing at an alarming rate, estimated as high as 50 square miles (or 32,000 acres) per year. National attention has focused on this problem. We are attempting to learn more about the causes of the losses (e.g., leveeing and channelizing the Mississippi, flow diversion, rising sea level, coastal subsidence,

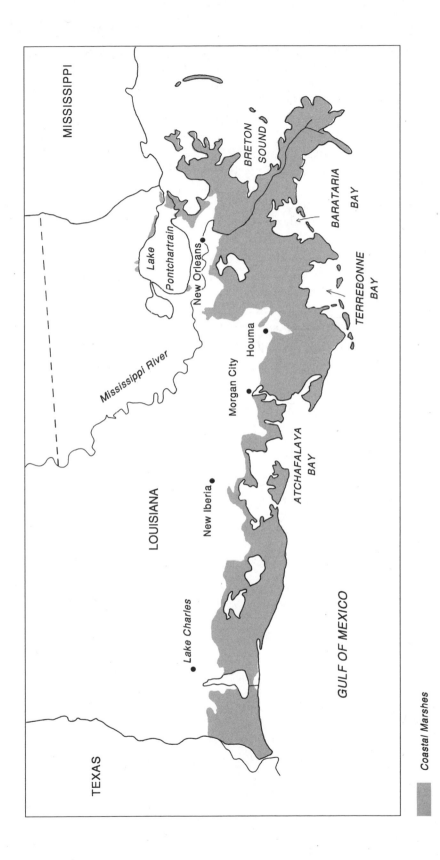

MISSISSIPPI

BRETON SOUND

BARATARIA BAY

Lake Pontchartrain

New Orleans

TERREBONNE BAY

Mississippi River

Houma

Morgan City

ATCHAFALAYA BAY

LOUISIANA

New Iberia

GULF OF MEXICO

Lake Charles

TEXAS

Coastal Marshes

canalization, and accelerated erosion) with the hope of developing management strategies to reduce future losses. More information on Louisiana's coastal marshes can be found in *Vegetation, Water, and Soil Characteristics of the Louisiana Coastal Region*, an Agricultural Experiment Station report from Louisiana State University at Baton Rouge.

Due to the prominence of coastal wetlands, they are most easily observed from the many roads linking coastal parishes with the interior (e.g., Route 82). The following areas, however, provide good opportunities to examine coastal marshes: (1) Sabine National Wildlife Refuge (Hackberry), (2) Rockefeller Refuge (Grand Chenier), and (3) Cypremont Point State Park (Franklin).

Texas

Texas has a high-energy coastline where numerous barrier islands have developed. Behind these islands and beaches (e.g., Matagorda Island, St. Joseph's Island, Mustang Island, and Padre Island) are shallow-water embayments (e.g., Galveston Bay, Matagorda Bay, San Antonio Bay, Aransas Bay, and Corpus Christi Bay). The Laguna Madre, a world-renowned hyperhaline lagoon, lies behind Padre Island. Coastal wetlands have developed on the landward side of the barrier islands and along the shores of the embayments as they do elsewhere, but due to local climate and infrequent flooding, they are high-salinity marshes with extensive salt flats. These marshes grade imperceptibly into coastal prairies dominated by Gulf cordgrass, which extend for miles inland along parts of the Texas coast. About 700,000 acres of coastal marshes and tidal flats may be present.

Some places to see coastal marshes are: (1) Laguna Atascosa National Wildlife Refuge (Rio Hondo), (2) Anahuac National Wildlife Refuge (Anahuac), (3) McFaddin Marsh and Texas Point National Wildlife Refuges (Sabine Pass), (4) Aransas National Wildlife Refuge (Austwell), (5) Galveston Island State Park (Galveston), and (6) Sea Rim State Park (Sabine Pass).

References

Adams, D. A. 1963. Factors influencing vascular plant zonation in North Carolina salt marshes. *Ecology* 44: 445–456.

Albert, R. 1975. Salt regulation in halophytes. *Oecologia* 21: 51–71.

Alexander, C. E., M. A. Broutman, and D. W. Field. 1986. *An Inventory of Coastal Wetlands of the U.S.A.* National Oceanic and Atmospheric Administration, Washington, DC.

Armstrong, N. E. 1987. *The Ecology of Open-bay Bottoms of Texas: A Community Profile.* U.S. Fish and Wildlife Service, Washington, DC. Biol. Rep. 85 (7.12).

Baker, M. F. 1926. *Florida Wild Flowers.* Macmillan, New York.

Boesch, D. F., D. Levin, D. Nummedal, and K. Bowles. 1983. *Subsidence in Coastal Louisiana: Causes, Rates, and Effects on Wetlands.* U.S. Fish and Wildlife Service, Washington, DC. FWS/OBS-83/26.

Britton, N. L., and C. F. Millspaugh. 1962. *The Bahama Flora.* Hafner, New York.

Brown, C. A. 1945. *Louisiana Trees and Shrubs.* Louisiana Forestry Commission, Baton Rouge.

Burk, C. J. 1962. The North Carolina Outer Banks: A floristic interpretation. *J. Mitchell Society* 78: 21–28.

Carlton, J. M. 1975. *A Guide to Common Florida Salt Marsh and Mangrove Vegetation.* Florida Department of Natural Resources, Marine Research Laboratory, St. Petersburg. Florida Marine Research Publications, No. 6.

Carlton, J. M. 1977. *A Survey of Selected Coastal Vegetation Communities of Florida.* Florida Department of Natural Resources, Marine Research Laboratory, St. Petersburg. Florida Marine Research Publications, No. 30.

Chabreck, R. H. 1972. *Vegetation, Water, and Soil Characteristics of the Louisiana Coastal Region.* Louisiana State University, Agricultural Experiment Station, Baton Rouge. Bulletin No. 664.

Cooper, A. W., and E. D. Waits. 1973. Vegetation types in an irregularly flooded salt marsh on the North Carolina Outer Banks. *J. Mitchell Society* 89: 78–91.

Correll, D. A., and H. B. Correll. 1972. *Aquatic and Wetland Plants of Southwestern United States.* U.S. Environmental Protection Agency, Washington, DC. Water Pollution Control Research Series 16030 DNL 01/72.

Cowardin, L. W., V. Carter, F. C. Golet, and E. T. LaRoe. 1979. *Classification of Wetlands and Deepwater Habitats of the United States.* U.S. Fish and Wildlife Service, Washington, DC. FWS/OBS-79/31.

Duncan, W. H., and M. B. Duncan. 1987. *The Smithsonian Guide to Seaside Plants of the Gulf and Atlantic Coasts.* Smithsonian Institution Press, Washington, DC.

Egler, F. E. 1952. Southeast saline Everglades vegetation, Florida, and its management. *Vegetatio* 3: 213–265.

Eleuterius, L. N. 1980. *An Illustrated Guide to Tidal Marsh Plants of Mississippi and Adjacent States.* Mississippi–Alabama Sea Grant Consortium, Ocean Springs, MS. MASG-77-039.

Eleuterius, L. N. 1972. The marshes of Mississippi. *Castanea* 37: 153–168.

Fernald, M. L. 1950. *Gray's Manual of Botany.* American Book Company, Boston.

Field, D. W., A. J. Reyer, P. V. Genovese, and D. B. Shearer. 1991. *Coastal Wetlands of the United States.* U.S. Department of Commerce, National Oceanic and Atmospheric Administration, Rockville, MD, in cooperation with the U.S. Department of Interior, Fish and Wildlife Service.

Gabriel, B. C., and A. A. de la Cruz. 1974. Species composition, standing stock, and net primary production of a salt marsh community in Mississippi. *Chesapeake Science* 15: 72–77.

Godfrey, R. K., and J. W. Wooten. 1979. *Aquatic and Wetland Plants of Southeastern United States: Monocotyledons.* University of Georgia Press, Athens.

Godfrey, R. K., and J. W. Wooten. 1981. *Aquatic and Wetland Plants of Southeastern United States: Dicotyledons.* University of Georgia Press, Athens.

Hanlon, R., F. Bayer, and G. Voss. 1975. *Guide to the Mangroves, Buttonwood, and Poisonous Shoreline Trees of Florida, the Gulf of Mexico, and the Caribbean Region.* University of Miami Sea Grant Program, Miami. Sea Grant Field Guide Series No. 3.

Hanlon, R., and G. Voss. 1975. *Guide to the Sea Grasses of Florida, the Gulf of Mexico, and the Caribbean Region.* University of Miami Sea Grant Program, Miami. Sea Grant Field Guide Series No. 4.

Harrar, E. S., and J. G. Harrar. 1962. *Guide to Southern Trees.* Dover, New York.

Higgins, E. A. T., R. D. Rappleye, and R. G. Brown. 1971. *The Flora and Ecology of Assateague Island.* University of Maryland, Agricultural Experiment Station, College Park. Bulletin A-172.

Hillestad, H. O., J. R. Bozeman, A. S. Johnson, C. W. Berisford, and J. J. Richardson. 1975. *The Ecology of the Cumberland Island National Seashore, Camden County, Georgia.* Georgia Marine Science Center, Skidaway Island. Technical Report Series No. 75-5.

Hitchcock, A. S. 1951. *Manual of the Grasses of the United States.* U.S. Department of Agriculture, Washington, DC. Misc. Publ. No. 200.

Jones, S. B., Jr. 1974. Mississippi Flora. I. Monocotyledon families with aquatic or wetland species. *Gulf Research Reports* 4(3): 357–379.

Kearney, T. H., Jr. 1900. The plant covering of Okracoke Island: A study in the ecology of the North Carolina strand vegetation. *U.S. National Herbarium* 5: 261–319.

Kurz, H., and K. Wagner. 1957. *Tidal Marshes of the Gulf and Atlantic Coasts of Northern Florida and Charleston, South Carolina.* Florida State University, Tallahassee. Florida State University Studies, No. 24.

Laessle, A. M. 1942. *The Plant Communities of the Welaka Area with Special Reference to Correlations between Soils and Vegetational Succession.* University of Florida, Gainesville. Biological Science Series, Vol. 4, No. 1.

Lellinger, D. B. 1985. *A Field Manual of the Ferns and Fern-Allies of the United States and Canada.* Smithsonian Institution Press, Washington, DC.

Linton, T. L. 1968. A description of the South Atlantic and Gulf coast marshes and estuaries. In J. D. Newson (ed.), *Proceedings of the Marsh and Estuary Management Symposium,* pp. 15–25. Louisiana State University, Baton Rouge.

Lugo, A. E., and S. C. Snedaker. 1974. The ecology of mangroves. *Ann. Rev. Ecol. Syst.* 5: 39–64.

Odum, W. E., C. C. McIvor, and T. J. Smith, III. 1982. *The Ecology of the Mangroves of South Florida: A Community Profile.* U.S. Fish and Wildlife Service, Washington, DC. FWS/OBS-81/24.

Odum, W. E., T. J. Smith, III, J. K. Hoover, and C. C. McIvor. 1984. *The Ecology of Tidal Freshwater Marshes of the United States East Coast: A Community Profile.* U.S. Fish and Wildlife Service, Washington, DC. FWS/OBS-83/17.

Olmstead, I. C., L. L. Loope, and R. P. Russell. 1981. *Vegetation of the Southern Coastal Region of Everglades National Park between Flamingo and Joe Bay.* Everglades National Park, South Florida Research Center, Homestead. Rep. T-620.

Radford, A. E., H. E. Ahles, and C. R. Bell. 1964. *Manual of the Vascular Flora of the Carolinas.* University of North Carolina Press, Chapel Hill.

Reed, P. B., Jr. 1988. *National List of Plant Species that Occur in Wetlands: Southeast (Region 2).* U.S. Fish and Wildlife Service, Washington, DC. Biol. Rep. 88 (26.2).

Reimold, R. J., and W. H. Queen (eds.). 1974. *Ecology of Halophytes.* Academic Press, New York.

Roe, C. E. 1987. *A Directory to North Carolina's Natural Areas.* N.C. Natural Heritage Foundation, Raleigh.

Silberhorn, G. M., G. M. Dawes, and T. A. Barnard, Jr. 1974. *Coastal Wetlands of Virginia: Interim Report No. 3*. Virginia Institute of Marine Science, Gloucester Point. Special Report in Applied Marine Science and Ocean Engineering No. 46.

Small, J. K. 1931. *Ferns of Florida*. Science Press, New York.

Stout, J. P. 1984. *The Ecology of Irregularly Flooded Salt Marshes of the Northeastern Gulf of Mexico: A Community Profile*. U.S. Fish and Wildlife Service, Washington, DC. Biol. Rep. 85 (7.1).

Tiner, R. W., Jr. 1977. *An Inventory of South Carolina's Coastal Marshes*. S.C. Marine Resources Center, Charleston. Technical Report No. 23.

Tiner, R. W., Jr. 1987. *A Field Guide to Coastal Wetland Plants of the Northeastern United States*. University of Massachusetts Press, Amherst.

Tiner, R. W., Jr. 1988. *Field Guide to Nontidal Wetland Identification*. Maryland Department of Natural Resources, Annapolis, and U.S. Fish and Wildlife Service, Newton Corner, MA. Cooperative publication.

Uhler, F. M., and N. Hotchkiss. 1968. Vegetation and its succession in marshes and estuaries along the South Atlantic and Gulf coasts. In J. D. Newson (ed.). *Proceedings of the Marsh and Estuary Management Symposium*, pp. 26–32. Louisiana State University, Baton Rouge.

U.S. Department of Commerce, National Oceanic and Atmospheric Administration. 1987. *Tide Tables 1988 High and Low Water Predictions: East Coast of North and South America Including Greenland*. National Ocean Service, Washington, DC.

Vince, S. W., S. R. Humphrey, and R. W. Simons. 1989. *The Ecology of Hydric Hammocks: A Community Profile*. U.S. Fish and Wildlife Service, Washington, DC. Biol. Rep. 85 (7.26).

Viosca, P., Jr. 1928. Louisiana wetlands and the value of their wildlife and fishery resources. *Ecology* 9: 216–229.

Wass, M. L., and T. D. Wright. 1969. *Coastal Wetlands of Virginia*. Virginia Institute of Marine Science, Gloucester Point. Special Report in Applied Marine Science and Ocean Engineering No. 10.

Wharton, C. H., W. M. Kitchens, A. C. Pendleton, and T. W. Sipe. 1982. *The Ecology of Bottomland Hardwood Swamps of the Southeast: A Community Profile*. U.S. Fish and Wildlife Service, Washington, DC. FWS/OBS-81/37.

Wiegert, R. G., and B. J. Freeman. 1990. *Tidal Salt Marshes of the Southeast Atlantic Coast: A Community Profile*. U.S. Fish and Wildlife Service, Washington, DC. Biol. Rep. 85 (7.29).

Wilson, K. A. 1962. *North Carolina's Wetlands: Their Distribution and Management*. N.C. Wildlife Resources Commission, Raleigh. Federal Aid in Wildlife Restoration Project W-6-R.

Zieman, J. C. 1982. *The Ecology of the Seagrasses of South Florida: A Community Profile*. U.S. Fish and Wildlife Service, Washington, DC. FWS/OBS-82/25.

Glossary

Achene. Small dry, hard, one-seeded nutlet.

Alternate (*leaves*). Arranged singly along the stem, alternating from one side to the other up the stem.

Angled (*stem*). Having distinct edges; three-angled (triangular in cross-section) and four-angled (square).

Annual. Plant living for only one year; propagates from seeds.

Anther. Distal end of a stamen where pollen is produced.

Appressed. Closely compacted together, as in an *appressed* inflorescence.

Aromatic. Sweet-smelling.

Arrowhead-shaped (*leaves*). Appearing like an arrowhead, triangular in shape.

Ascending. Rising upward and somewhat spreading, as in an *ascending* inflorescence.

Awn. Bristle-shaped appendage.

Axil. Angle formed by a leaf or branch with the stem.

Axillary. Located in an axil.

Axis. The central portion of an organ-bearing structure, as in an inflorescence of a grass the *axis* bears the spikelets.

Basal (*leaves*). Arising directly from the roots; may ascend along stem as sheaths and appear alternately arranged, as in cattails.

Beak. Long, thickened point or tip.

Berry. Fleshy or pulpy fruit.

Blade. Flattened leaf.

Bract. Leaflike or modified appendage subtending a flower or belonging to an inflorescence.

Bristle. Long stiff, hairlike structure.

Bud. Unexpanded flower or leaf.

Callous. Fleshy, thickened tissue, as in the tips of leaf teeth in certain plants.

Calyx. Outermost parts of a flower; refers to the sepals, which are usually green but sometimes colored and petallike.

Capsule. Dry fruit composed of two or more cells or chambers.

Catkin. Scaly spike of inconspicuous flowers lacking petals, usually male or female.

Cell. One of the chambers of a capsule.

Channelled. Having distinct grooves or ridges.

Cilia. Fringing or marginal hairs.

Clasping (*leaves*). Closely surrounding the stem and attached directly without stalk.

Compound (*leaves*). Divided into two or more distinct, separate parts (leaflets).

Coniferous. Cone-bearing.

Corm. Enlarged fleshy base of stem; bulblike.

Corolla. Petals of a flower.

Corymb. Somewhat flat-topped inflorescence with outer flowers blooming first.

Cyme. Flowering inflorescence with innermost or terminal flowers blooming first.

Deciduous. Not persistent, dropping off plant after completing its function, as with *deciduous* leaves in fall.

Decumbent. Reclining or prostrate at base, with the upper part erect or ascending, as in *decumbent* stems.

Dioecious. Having two types of flowers (male and female) borne on separate plants.

Disk. Tubular flower forming the central head of composites or asters.

Dissected. Deeply divided, often into threadlike parts, as in *dissected* leaves.

Drupe. Fleshy or pulpy fruit having a single stone or pit.

Emergent. Herbaceous (nonwoody) plant standing erect.

Entire (*leaves*). Having smooth margins, without teeth.

Epiphyte. Type of nonparasitic plant growing on other plants; commonly called "air plant."

Evergreen. Persistent, as in *evergreen* leaves that remain on plant through winter.

Filament. Basal part of a stamen that supports the anther.

Fleshy. Soft, thickened tissue; succulent.

Follicle. Dry fruit that opens along one line or suture, as in milkweed and bean pods.

Frond. Leaf of a fern.

Gland. Secreting structure or organ.

Glandular. Bearing glands.

Glomerule. Compact somewhat roundish flowering head.

Glume. Thin bract at the base of a grass spikelet.

Grain. Fruit of certain grasses.

Head. Dense cluster of sessile or nearly sessile flowers, characteristic of composites or asters.

Herbaceous. Nonwoody.

Hood. Erect, outermost "petals" of milkweed flowers.

Horn. Erect, inner tubular structure of milkweed flowers.

Inflorescence. Flowering part of a plant.

Internode. Portion of a stem between two nodes.

Irregular (flower). Similar parts (e.g., petals) differing in size and/or shape.

Irregularly flooded. Flooded by tides less than once daily.

Jointed (stem). Having obvious nodes.

Lance-shaped (leaves). Appearing as the head of a lance, several times longer than wide, broadest just above the base, tapering to a tip.

Lateral. Borne on the sides of a plant.

Lemma. Lower of two bracts enclosing the flower of a grass.

Lenticel. Corky spot or line, sometimes raised, on the bark of many trees and shrubs.

Ligule. Membranous or hairy structure at the junction of the leaf blade and the leaf sheath in grasses.

Linear. Narrow and elongate, several to many times longer than wide.

Lip. Upper and lower parts of certain tubular flowers.

Lobe. Indented part of leaf or flower, not divided into distinct and separate parts but still interconnected to similar parts of leaf or flower (e.g., petal).

Midrib. Central, prominent rib or main vein of a leaf, usually in center of leaf.

Midvein. Middle vein of a leaf.

Monoecious. Bearing both male and female flowers.

Mucronate. Abruptly sharp-pointed.

Nerve. Vein of a leaf, usually the more prominent ones.

Node. Point of a stem where leaves and branches are produced.

Nutlet. A small, dry, hard fruit.

Oblong (leaves). Longer than wide, with nearly parallel sides.

Ocrea. Tubular stipule, in smartweeds becoming fibrous.

Opposite (leaves). Arranged in pairs along the stem.

Orifice. Opening of a leaf sheath along the stem.

Oval. Broadly egg-shaped, widest in the middle and tapering to the ends.

Ovary. Part of a pistil containing the seeds.

Palea. Upper of two bracts enclosing the flower of a grass.

Panicle. Much branched flowering inflorescence.

Panne. Shallow depression within irregularly flooded salt marshes.

Papillose. Bearing short, minute wartlike or nipplelike structures.

Pedicel. Stalk of a single flower in a cluster.

Peduncle. Primary stalk of a flowering cluster or single flower.

Peltate (leaves). Having stalks attached to center of leaf from beneath.

Perennial. Plant living for many years, usually supported by underground parts, e.g., rhizomes, corms, tubers, or bulbs.

Perigynium. Inflated sac enclosing the seed of a sedge.

Persistent. Remaining on plant after function ceases, as in *persistent* fruits.

Petiole. Stalk of a leaf.

Phyllode. Leaflike flattened petiole that functions as a leaf.

Pinnate (leaves). Divided into leaflets that are oppositely arranged.

Pistil. Seed-bearing structure of a flower, usually consisting of an ovary, stigma, and style.

Pith. Soft, fleshy, or spongy center of a stem.

Plano-convex. Flattened but somewhat curved.

Pneumatophore. Erect aerial growth from underground roots, functioning to improve aeration in certain mangroves.

Pod. Dry fruit capsule.

Prickly. Bearing small spines.

Prostrate. Lying flat on the ground.

Raceme. Spikelike inflorescence with stalked flowers.

Rachis. Main axis of a spike, branching inflorescence, or compound leaf.

Ranks. Number of rows of organs, such as leaves, along a stem.

Ray. Outer flower of the flowering head of composites or asters, often petallike.

Recurved. Curved downward.

Regular (flower). Similar flower parts of the same size and shape, radially symmetrical.

Regularly flooded. Flooded by tides at least once a day.

Rhizome. Underground part of a stem, usually horizontal and rooting at nodes and producing erect stems.

Runner. Prostrate, slender aboveground stem producing new plants at nodes.

Samara. Winged dry fruit bearing one seed.

Scale. Modified leaf or thin flattened structure.

Scape. Naked flowering stalk arising directly from roots.

Sepal. Outermost part of a flower, usually green but sometimes colored and petallike.

Septa. Partitions.

Sessile. Without stalks, as in *sessile* leaves that are attached directly to the stem without stalks.

Sheath. Tubular envelope surrounding the stem, as in leaf *sheaths* of grasses and sedges.

Shrub. Erect, woody plant less than 20 feet tall, usually with multiple stems but also including saplings of tree species.

Simple (leaves). Not divided into separate parts; leaf blade continuous.

Sinus. Space between two lobes.

Sori. Cluster of sporangia (fruit dots) of ferns.

Spadix. Fleshy spike.

Spathe. Large bract or pair of bracts enclosing an inflorescence.

Spatulate. Spoon-shaped.

Spike. Simple, unbranched inflorescence composed of a central axis with sessile or nearly sessile flowers.

Spikelet. A subdivision of a spike.

Spine. Sharp-pointed outgrowth of stem.

Sporangia. Spore cases of ferns, horsetails, and quillworts.

Spore. Reproductive structure of ferns, horsetails, and quillworts.

Spur. Hollow, tubular extension of a flower, usually bearing nectar.

Stamen. Pollen-bearing part of a flower.

Stigma. Part of a pistil receiving and germinating pollen.

Stipules. Pair of appendages at the base of a leaf stalk or on each side of its attachments to the stem.

Stolon. Prostrate, slender aboveground stem producing new plants at nodes.

Style. Part of a pistil connecting the stigma with the ovary.

Submerged. Underwater.

Subtended. Lying below.

Succulent. Fleshy.

Sword-shaped (leaves). Appearing bayonet-shaped, flattened and tapering to a sharp-pointed tip.

Synonym. Previous taxonomic or scientific name.

Taproot. Prominent, deep-penetrating root.

Tidal. Subject to influence of ocean-driven tides.

Thallus. Plants without clear separation of leaves and stems, as in duckweeds.

Translucent. Opaque, nearly see-through.

Tree. Woody plant 20 feet or taller with a single main stem (trunk).

Tuber. Short, thickened, usually underground stem, having buds or eyes and storing food.

Tubercle. Small thickened protuberance.

Turbinate. Top-shaped or inversely cone-shaped.

Twining. Climbing by wrapping around another plant or other support.

Umbel. Branched inflorescence with flowering stalks arising from a single point.

Unisexual. Bearing male or female parts but not both, as in male or female flowers.

Valve. Piece of an open capsule.

Vascular. Having vessels or ducts.

Veins. Threads of vascular tissue in a leaf.

Whorl. Three or more organs arranged in a circle around the stem, as in *whorled* leaves.

Wing. Flattened expansion of an organ, as the continuation of a leaf as a *wing* along the stem.

Index

Page references in boldface refer to descriptions of major illustrated species.

CONVERSION TABLE

English Units	Metric Equivalents
1/25 inch	1 millimeter
1/5 inch	5 millimeters
1/4 inch	6 millimeters
1/2 inch	12 millimeters
1 inch	2.5 centimeters
1 foot	30 centimeters
3.3 feet	1 meter
10 feet	3 meters